— PRAISE FOR —
BRINGING BACK THE BEAVER

'*Bringing Back the Beaver* is a hilarious, eccentric and magnificent account of a struggle against bureaucracy, pigheadedness and sheer human irrationality, in order to reintroduce a species crucial to the health of our ecosystems. Derek Gow is an extraordinary character, whose writing is as colourful and dynamic as he is. He has done more to restore our missing fauna than anyone else in Britain. This is his astonishing story of what it takes.'   —GEORGE MONBIOT

'It is wonderful to see that beavers are now officially back on the list of native species, having been absent for so long . . . far too long! They are without doubt one of the world's best natural engineers with their ability to create new wetland habitats, which then become home to many other creatures who need a "watery" landscape in which to live and survive. This is a win–win situation for wildlife . . . providing of course, reintroductions are carried out with the blessing of the surrounding landowners and the local community.'

—DAME JUDI DENCH

'Derek Gow might be the most colorful character in all of Beaverdom – a wry, profane truth teller who is equal parts yeoman farmer, historical ecologist, and pirate. *Bringing Back the Beaver* is a swashbuckling saga whose banks overflow with semi-legal hijinks, ribald jokes and hard-won scientific insight. It's also a keen lament for the nature we've stamped out, and an impassioned ode to rewilding that will inspire Beaver Believers in Britain and beyond.'

—BEN GOLDFARB, author of *Eager*

'Derek Gow has produced a fearless work – friends and foe alike feel the witty sting of his barbed tongue. Do not mess with him! Above all though, this book is important – an accessible insight into what can be achieved when rewildling is taken out of its gilded cage and given a pragmatic kick in the backside. Beavers are back – thanks in large part to Derek. Read the story, learn why beavers are vital to our ecology and, above all, have fun.'

—HUGH WARWICK, author and ecologist

'Beautifully written, passionate, blunt, uncompromising and very, very funny; not many books make me laugh out loud, but this one had me in stitches. Derek Gow is a man after my own heart, with no interest in diplomacy, but with single-minded determination and a clear vision: to bring beavers back, and to bring nature back, to Britain.'

—DAVE GOULSON, author of *A Sting in the Tale* and *The Garden Jungle*

'*Bringing Back the Beaver* is brilliant: passionate, humorous, inspiring and full of hope. It's not just about beavers – it's about humans, too, and how, when we really set our minds to it, we *can* make a difference. I already loved the idea of beavers being brought back to the UK, but by the time I'd finished reading Derek's book, I was ready to chain myself to the railings outside 10 Downing Street to help make it happen.'

—BRIGIT STRAWBRIDGE HOWARD, author of *Dancing with Bees*

'A wonderfully entertaining account of the battle to return the Eurasian beaver to the British countryside from the "Beaver Man" himself – Derek "bloody" Gow. No one has done more to promote their return to Britain and it is entirely fitting that it is he who reports on the sorry – and sadly still ongoing – catalogue of endless bureaucratic procrastination and adversarial clutching of straws that has prevented us from embracing the most blindingly obvious catchment restoration tool known to man.'

—PROFESSOR ALASTAIR DRIVER, director, Rewilding Britain

'Gripping, illuminating, informative, amusing and inspiring in equal parts – I loved it. *Bringing Back the Beaver* is not only a ripping yarn with interesting and funny characters, tension and drama, it is an important tale of tenacity, the tenacity we need much more of to repair our depleted land.'

—LUCY JONES, author of *Losing Eden* and *Foxes Unearthed*

'In our age of massive wildlife depletion, British lovers of nature have long been crying out for major change. Cometh the hour, cometh the man and his beast. In his fact-filled, funny and irrefutably argued book, Derek Gow sets out a powerful case for the return of a creature that will transform our wetlands for the better. I urge you to read his book. It will make "beaver believers" of us all.'

—MARK COCKER, author and naturalist

'A tail-slapping yarn, interwoven with emoji-faced farmers, blundering officials, gothic scientists and his own unwavering mission to restore the beaver, Britain's bringer of life. Derek is a naturalist with a strong case of beaver fever shared effusively with equal doses of charm, belligerence and artistic genius. May his beautifully illustrated story bust the dams of progress in these hot and stormy times.'

—JAMES WALLACE, director, Beaver Trust

'In characteristic punchy style, Derek Gow has given us the full saga of beaver reintroduction to the UK. Throughout a shameful tale of ministerial ignorance and dithering, Derek has charted his own remarkable role in ensuring this vital keystone species is restored to our wetlands. The book is a triumph of determination over bureaucratic obfuscation, an essential read for everyone interested in nature conservation.'

—SIR JOHN LISTER-KAYE, OBE, naturalist and nature writer

'Like his subject, Derek Gow is industrious – woodily, wetly – and deeply committed to his project. To have beavers back in Britain gives all of nature – us included – a second chance at happier times. Gow tells us this with the clear-sighted and good-humoured energy and purposefulness of the animals he has come to love as well as admire. We all should be converted to his cause.'

—TIM DEE, author of *Landfill* and *Greenery*

'The fun of Derek Gow's book comes from the way it weaves together the reintroduction of beavers to Britain with his own rollicking life in the cause of ecological restoration. Its "upshot," to borrow a word from Aldo Leopold, is Gow's depiction, at once comprehensive and specific, of wetlands' importance in restoring what had become a simplified, dwindling countryside. In his highly engaging way, he conveys beavers' essential role in the landscape's biological health and diversity: ". . . it needs them and without them would fade swiftly away."'

—JOHN ELDER, author of *Reading the Mountains of Home*

'Derek Gow assails officialdom in this unsparing and funny memoir, recounting the long struggle to bring wild beavers back to Britain. A witty and passionate guide, he'll have you rooting for the rodents from page one, but the real stars of the story may be the colorful characters working on their behalf, including the author himself. Highly readable and highly recommended.'

—THOR HANSON, author of *Buzz*

Bringing Back the Beaver

The Story of One Man's Quest to Rewild Britain's Waterways

Derek Gow

Foreword by Isabella Tree

Chelsea Green Publishing
White River Junction, Vermont
London, UK

Project Manager: Patricia Stone
Commissioning Editor: Jonathan Rae
Developmental Editor: Michael Metivier
Copy Editor: Marisa Crumb
Proofreader: Angela Boyle
Indexer: Shana Milkie
Designer: Melissa Jacobson

Printed in the United States of America.
First printing August 2020.
10 9 8 7 6 5 4 3 2 1   20 21 22 23 24

Our Commitment to Green Publishing

Chelsea Green sees publishing as a tool for cultural change and ecological stewardship. We strive to align our book manufacturing practices with our editorial mission and to reduce the impact of our business enterprise in the environment. We print our books and catalogs on chlorine-free recycled paper, using vegetable-based inks whenever possible. This book may cost slightly more because it was printed on paper that contains recycled fiber, and we hope you'll agree that it's worth it. *Bringing Back the Beaver* was printed on paper supplied by Sheridan that is made of recycled materials and other controlled sources.

Library of Congress Cataloging-in-Publication Data
Names: Gow, Derek, author.
Title: Bringing back the beaver : the story of one man's quest to rewild Britain's waterways / Derek Gow.
Description: White River Junction, Vermont : Chelsea Green Publishing, 2020. | Includes bibliographical references and index.
Identifiers: LCCN 2020026185 (print) | LCCN 2020026186 (ebook) | ISBN 9781603589963 (hardcover) | ISBN 9781603589970 (ebook) | ISBN 9781603589987 (audio)
Subjects: LCSH: Beavers—Great Britain—Reintroduction.
Classification: LCC QL737.R632 G695 2020 (print) | LCC QL737.R632 (ebook) | DDC 599.370941—dc23
LC record available at https://lccn.loc.gov/2020026185
LC ebook record available at https://lccn.loc.gov/2020026186

Chelsea Green Publishing
85 North Main Street, Suite 120
White River Junction, Vermont USA

Somerset House
London, UK

www.chelseagreen.com

To Gerhard Schwab,

Who has done so much in his lifetime to forge a new type of relationship with the beavers he loves based on respect and understanding.

— CONTENTS —

— FOREWORD —

W E ARE ONLY JUST BEGINNING TO UNDERSTAND the extent of the beaver's role as a 'keystone species' – a creature that has a disproportion-ately large effect on its environment. Like the keystone of an arch, biological structures, entire ecosystems, depend on it. No other creature it seems – other than perhaps elephants and humans – has such a profound and dramatic impact on the landscape. The beaver is an ecosystem engineer, architect of watery kingdoms and riparian habitats teeming with life, restorer of natural hydrology, creator – even – of soil itself.

Our world without them, since we hunted them to the verge of extinction for their sleek, water-repellent fur, has been a poorer place. And we have grown used to this depleted, simplified, beaver-less world. In the UK, where beavers disappeared centuries earlier than in Europe, they have been relegated to the fairy tales of Narnia where C.S. Lewis, to his eternal disgrace, has these obligate herbivores eating fish – misinformation that, like the man-eating wolf in Little Red Riding Hood, refuses to lie down.

The idea that the long-lost beaver should be returned to our countryside stirs up all sorts of feelings of discomfort and distrust. The very fact that its impact is so great engenders concern, particularly amongst the inhabitants of the British Isles, which lost most of its large, charismatic, native mammals long, long ago. In our tight, micro-managed, densely populated modern landscape, it is hard to envisage how the creative dynamism of the beaver will fit in. How will it affect our river systems? Will it cause flooding to farmland, crops, property? Is today's environment appropriate for an animal that has been absent since the Middle Ages?

It takes knowledge, imagination and pragmatism to assuage these anxieties, to bring the fears and fantasies back down to earth and shift the mindset of the intransigent. All these qualities Derek Gow has in spades. But he is not your archetypal nature conservationist. Not for him the diplomatic chess-game, the ticking of boxes, the subtle play of compromise, of deferential negotiation with policy makers, NGOs and the powers-that-be. His story, told here with the vim of the crusader and the no-holds-barred stab of the stand-up comedian, is about the nonsensical levels of bureaucracy, the petty feuds, the miasma of misinformation and the overbearing caution that stood in his way as he and a few dedicated beaver exponents and reintroduction specialists, most of whom we meet in this book, tried – and ultimately succeeded – bringing this charismatic creature back to Britain.

What Derek appreciates, more than most, is the urgency of the situation. There is no time, he rages, for prevarication and yet more environmental impact studies.

Foreword

Research in Europe and the United States has shown, incontrovertibly, the beneficial effects that beavers have on the environment. Beavers are native to Britain. They populate our archaeological records, our parish registers, our place names, the parliamentary acts that required us to treat them as vermin. Nature in the UK needs all the help it can get, and the beaver is key to ecological recovery, the desperately needed shot in the arm.

To many of the government officials, farmers, anglers, landowners and conservationists with whom he has locked horns, Derek Gow is a 'one-man wrecking ball', a 'cowboy operating outside the law', a 'complete pain in the arse'. Some of his friends might say the same. But most also appreciate his candour and bull-headed determination, his ability to cut through the nonsense, to get things done. For environmental campaigners who share his vision but lack his gumption, he is a potent megaphone. Put Derek on stage and he will often have his audience weeping with emotion and whooping encouragement. His passionate eloquence stems, above all, from a fundamental need to communicate and inspire, to get the ball rolling.

Like many mavericks he is, largely, an autodidact – an independent thinker who has acquired his knowledge of wildlife and animal husbandry through voracious reading, hands-on experience and astonishingly astute powers of observation. He grew up in the Scottish Borders in the tiny village of Broughton in the upper valley of the River Tweed. Having left school at seventeen (he was tempted by art college but pragmatically opted for a surer way of making a living), he spent six years as a livestock auctioneer in Edinburgh, building on his childhood hobby of raising a small flock of sheep for market. His eye for a good animal and his ability to communicate with the market drovers, dealers and punters landed him a job in environmental education at Palacerigg Country Park on a seasonal basis, as a ranger. In 1990 he attended a summer school on captive breeding endangered species at Gerald Durrell's zoo on Jersey. He'd read every one of Durrell's books. He left the programme a changed man – committed, as he puts it, to 'the salvation of wild creatures'.

Since then, Derek has bred and housed everything from wildcats, white storks, European field hamsters and harvest mice to night herons, stoats, water shrews and polecats at his farm in Devon. To date, he has released over 25,000 water voles – another keystone species, whose population in the UK has crashed 95% according to a survey conducted in 1992 – into twenty-five restored areas of wetland from Aberfoyle in Scotland to the River Meon in Hampshire. It was his observations of the water vole and its ecosystem that drew him to the beaver. Left to their own devices, he noticed, restored wetlands would sooner or later silt up, becoming engulfed by reeds and, eventually, scrub and trees – requiring enormous human effort and management to keep them open. What would have kept these wetlands open and dynamic in the past, before humans? Soon, he was interpreting the landscape in a totally different way, recognising beneath the veneer of modern human activity the ghosts of ancient beaver dams and the silted pans they'd left behind.

His first encounter with beavers in the wild in Poland opened up a world of biological connectivity. Amongst the swamps generated by gigantic active beaver dams, he saw floating islands carpeted with orchids and wolves bouncing roe deer into the bogs to trap them. In Bavaria he watched a sand lizard basking on the edge of a beaver dam. She slid off into the water at his approach, her tail propelling her downwards like a marine iguana, and popped up onto a beaver-gnawed log some distance away to bask again. The Bavarian ecologist with him said this was common. The sand lizards (according to all guide books, a definitively dry heathland or sand-dune reptile) were – like so many other unexpected species in these swamps, including red-backed shrikes – attracted to the insects swarming around the water-sodden beaver logs. If he hadn't seen it for himself, Derek could never have imagined it. It led him to appreciate how little we know. Our modern landscape is generally so changed, so desperately impoverished, we may be observing species not in their preferred habitat at all, but where they are simply clinging on for dear life. Given richer opportunities – better habitat and resources – wildlife might behave in entirely different ways. Only by releasing man's constraints on nature, by letting the genie out of the bottle, can we ever know what might be.

In many ways, Derek appreciates, this is a leap of faith, but it is one, he feels, that is desperately worth making.

For this consummate doer, this man of action, letting go, surrendering the driving seat to keystone species like the beaver, is the best thing we can possibly do for nature. He knows the battle is not yet won. Beavers might be back in starter populations in Scotland, on trial on the River Otter in Devon and in licensed, enclosed release sites elsewhere in England but their presence here is still tenuous. Perhaps Derek's greatest contribution yet to the future of the beaver in Britain may be his ability to break down barriers, bash heads together, inform, cajole, inspire and excite, as he does so convincingly in this refreshingly candid book.

Isabella Tree
May 2020

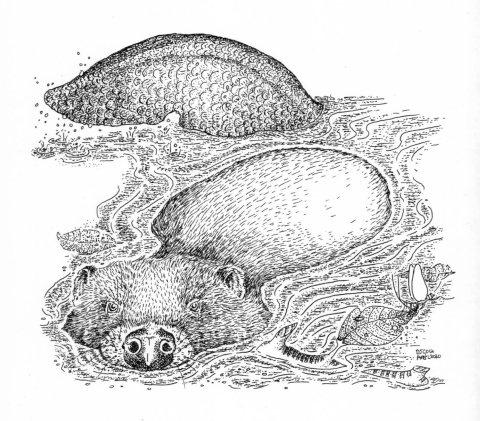

Prologue

A DAMP, DARK HEAD BROKE THE SURFACE OF THE amber pool.

We looked at the beaver and it looked back at us. Droplets of water dangled delicately from its starlet lashes. It blinked before sighing gently and disappearing with a swirl back into the tea-dark depths just as silently as it had surfaced.

The fact that it was on the wrong side of the fence was, frankly, a surprise. The implications were not lost on either of us. 'Oh dear,' said Toby, 'that's rather trying. Who would have thought it possible?' As I looked at the tangled strands of decaying sheep netting descending down into the channel, it occurred to me that the beaver's escape was more inevitable than impossible. Toby's enclosure – which had looked so emphatically solid on land – was, at least underwater, based more on faith than fact.

In the moments that followed, as we digested the potential fallout of what we had just witnessed, elation was not an emotion that sprang to the fore. While both of us were completely committed to reintroducing beavers back into the British countryside, we knew that an actual release, however accidental, would play badly in the toxic political atmosphere of the times when the mere theory was being roundly opposed by bewhiskered, tweedy opponents in fortified castles of great antiquity. As the nearest beaver traps to hand were approximately 500 miles away in Kent, only one option remained: to go home before the police arrived.

When they finally did a decade later, beavers were well established on the River Tay. No one knew exactly where they had come from as Toby's had been recaptured long before. Information was misty. Grey and intangible. Rumors of escapes from other enclosures in oxbows on estates where ex-cavalry officers presumed 'no need for fencing' as beavers were known to 'not walk overland' might have provided some founders.

In 1994, Scottish Natural Heritage (SNH) – the Scottish Government's nature conservation agency – decided to reintroduce beavers. The Eurasian species (*Castor fiber)* had, after hitting its lowest ebb of perhaps around 400 individuals at the beginning of the twentieth century in western Europe, been restored back into virtually every range state where it had formerly existed. When suitable habitats remained, and if sufficient beavers were available, reintroductions had not been hard to accomplish. Britain was pretty much the last in line. Based as it was on logic, the case advanced by SNH was, they believed, going to be simple to execute.

They were wrong.

Prologue

Anglers blustered that their dams would act as barriers that migratory game fish would be unable to negotiate, and that in any case the beavers were just another unwanted predator. Regardless of the fact that beavers are vegetarians, some still maintain an opposite belief to this day. The private forestry sector was also unimpressed. It was stuffed full with investors in eastern European countries whose third-world infrastructure of poorly designed drainage systems, installed in the aftermath of World War II to enable the creation of conifer plantations, provided recovering beaver populations with the opportunity to cheerfully readapt them back into the swamplands they once were. The gigantic furry rats were denting their profit projections.

Grumpy farmers said what they always say – NO! – and persuaded political chums who liked to drink whiskey, bekilted in their company at the Royal Highland Show that their views alone should prevail.*

Civil servants demurred or made mythic obstacles, while occasionally expressing languid commitment.

---

* Years ago, one farmer who blessed me with the time required to explain why we wanted to introduce beavers into his wetlands said, 'Well, Derek, I know nothing about this. Never thought about it at all. It makes as much sense to me as if you had come in here and said that you wished to put a great white shark in the reservoir, a giraffe on the hill in the far distance and a troop of dog-faced baboons in the trees below my house. Now bugger off.'

3

Politicians shimmied with whichever wind blew.

But people liked the prospect. When initial public consultations demonstrated a seam of enthusiasm for the idea, more were commissioned. When those demonstrated that support was rising, opponents complained that views of this sort did not matter. That their advocates were 'townies' who understood nothing. Although it was 'townies' taxes that underwrote their lifestyles; the wishes of ordinary people were pretty much ignored. The debate generated nothing more than hot air for more than fifteen years.

Until a man named Nevin changed all that.

———

In the late 1990s, the game-farming owner of a Perthshire wildlife park obtained Eurasian beavers. Nevin was tough, a sharply sarcastic individual, generally resplendent in dung-coloured apparel of big socks, plus-four trousers, battered brown brogues and a sordid flat cap stained thoroughly with the blood of the roe deer he stalked at first light with his Belgian hunting clients. When he was young, tufts of red hair sprouted readily from his ears. As he got older, they went grey. Although some might have considered him to be an effective hybrid between one of Queen Victoria's more martial highland retainers and a squirrel, no one was ever brave enough to say so to his face. He hated authority. He hated birds of prey. He hated vegans, and although not generally keen on even his paying customers, he did enjoy insulting them roundly if they complained about any aspect of his visitor centre.

Nevin's wildlife park contained a broad range of creatures. Sitatunga antelope from Africa, chamois from the Alps, Asian short-clawed otters, white storks and kookaburras rubbed along without theme or reason. Some inhabitants, such as the large crested porcupines, escaped to consume vast quantities of vegetables in the neat allotments of the genteel, retired middle classes in the well-heeled village of Comrie. Others, such as his ever-expanding colony of North American prairie dogs, burrowed to freedom from their hilltop location to eventually emerge in the farmlands below. There in the soft, friable soils of the pastures and the rich, ripe fields of corn, their engineering abilities – untrammelled by rocks – could be fully expressed. As rural dwellers discovered, gazing astounded when gargantuan tractors plunged funnel deep into the ground, they rapidly built their own Elysium. When the authorities complained and Nevin told them to 'fuck off', they promptly complied. In an age when zoos were meant to have a serious conservation purpose, Nevin's had none. His operation was simply there to make money.

Nevin's beavers were black.* They came from the state farm at Popielno in Poland and arrived at his wildlife

---

* As are about a third of the eastern European populations. Although this form was at one time more widespread in western Europe, in modern times, they are virtually unknown.

centre in early 2001 once their period of statutory quarantine at Blackpool Zoo was complete.

Announcing their arrival, Nevin put out a press release stating that he had imported beavers and was releasing them into a secret location in Scotland the following day. The media reception this received was exceptional, and most of the national papers, TV and radio covered it as either a good or bad news story depending on their editorial perspective. The CEO of SNH was called in astounded from his golf course on the Sunday morning the story was launched, and as he knew nothing whatsoever about what was happening, he issued an impressive tirade of completely pointless threats. Although it made for a good story, the 'release' was a simple zoo-to-zoo movement, which, when the heat of the moment expired, left a delighted Nevin with no case to answer and considerable, positive, free PR. Visitors poured into his wildlife centre where real creatures and oddities of his eccentric imaginings rubbed shoulders. A sign on an enclosure that read WATER OTTER encouraged perplexed children to pull on a rope that raised a sliding door on a chicken coop to reveal a small red kettle with OTTER THAN IT WAS TO BEGIN WITH painted on its side.

After a few months went by, the beavers were settled into a small but robustly fenced enclosure where an internal mains hotwire packed a punch that would have stunned a tyrannosaur trying to breach its fortified bounds. It was safe. Very secure. All was well until one day when the keeper in charge of the beavers straddled the metal stile into their pen with a large bucket of carrots and discovered instantly, suddenly and very painfully that by placing a branch on the electric wire running around the internal fence the beavers had connected it to the stile. Her carrots went skywards. She screamed, and after hobbling over to switch off the fence went to bed weeping without telling anyone else.

When she came back to work three days later, the carrots inside the enclosure were still there; nothing had touched them and the beavers had gone.

Nevin replaced them with others from a zoo on the Isle of Wight and for a time no one was any the wiser. He hoped the escapees were dead. They were not.

One of the first to twig to the prospect that Scotland may have obtained a free-living beaver population by accident rather than design was Hugh Chalmers, who at that time (2003) was the conservation forester for the Borders Forest Trust. Hugh, a genial, bearded, ginger sort of a chap, was canoeing up the River Isla in Glenfarg. The day was mild. The weather was clear. The birds were singing. Relaxed by the rhythm of his paddle strokes and lulled by the sounds of nature, Hugh, who was familiar with beaver-generated landscapes elsewhere in Europe, was contemplating just how suitable the willow wood-lands on the river's banks would one day be for beavers when a beaver swam right past him. He flailed in a fug of excitement for his mobile phone in its waterproof pouch with which, when extracted, he called me.

'I don't know, they are not mine', I replied when I had finally made sense of Hugh's breathless panting.

'Well it's definitely a beaver with a big paddle tail, which it slapped on the water when the canoe went by. It climbed up the bank and cut down a tree! We circled. We saw it. It's a beaver! The only odd thing about it was that it was black!'

With a sinking feeling I called Nevin who, after denying at length that he was in any way responsible, eventually admitted that he was.

After this the beavers disappeared again from view and it was not until 2006 – when numerous mature deciduous trees lining the pools of a fish farm near Perth were discovered pointing up the wrong way – that any other tangible evidence of their presence emerged. Further investigations revealed suspicious piles of wood-chips around their 'pencil-sharpened' stumps, and this, together with removed side branches, distinctive tooth marks and well-worn foraging paths leading from where the trees used to be back into the water, indicated that beavers were probably responsible. The nature conser-vation authorities were bemused and although a slow awareness of beavers on the Tay was growing – one farmer was astounded to discover that the mink on his burn had developed the ability to build dams out of potatoes – did nothing in response.

———

Toby and his brilliantly combative wife, Lavinia, are the modern lairds of a crumbling sixteenth-century castle in

Perthshire, and longstanding advocates of beaver reintroduction. Toby, despite his Scottish roots, has a prim English
accent and looks generally pert in a beret. Provided with
an old black bike and a basket, some strings of onions and
a striped shirt, he would without issue metamorphose
into something quite French. Lavinia is made of sterner
stuff. A dogged activist and Green Party campaigner, she
is a formidable, female battleship. Although she's quite
cuddly at times, I have on occasion been on the receiving
end of a few of her broadsides and as a result am not
keen to arouse her ire further. In general, she's not a
great fan of my jokes, but she will hopefully warm to
the description I have provided of her and greet me with
cake rather than a cleaver when next we meet.

After Toby and I discovered their temporary escape,
the beavers on their estate continued to prosper, and
over the years he obtained more from several sources.
Another import of black ones from Poland were eventually provided with cinnamon mates from Bavaria,
and while a pair of large, blond (Viking?) beavers were
received from Norway, these did not live long enough
to produce any offspring. All were legally enclosed
behind fences to create 'demonstration' wetlands where
scientists, farmers and fishermen with open minds could
mingle with general visitors to gain some understanding
of the complexity of beaver-generated landscapes.

As I hope this book makes clear, this knowledge is critical.

CHAPTER ONE

The Salvation
of St Felix

A CCORDING TO AN OLD FOLK TALE, WHEN A SHIP carrying St Felix of Burgundy was wrecked in a storm on the River Babingley in Norfolk in 615 CE, the saint was saved from drowning by a colony of beavers. In gratitude he consecrated the chief beaver as a bishop. The village, which is now abandoned, records this event on its signpost where a large beaver wearing a bishop's mitre administers to another more junior candidate.

But beavers have no patron saint, and while others have blessed their utility, once Felix was saved he never looked back.

Although in modern times humans and beavers have as species a relationship of great interwoven complexity, like Felix we have long forgotten their abilities. In western European landscapes that we've adapted utterly to suit our multifarious needs, beavers until very recent times have been absent. We killed them. Nearly all. The prospect of us tolerating them building dams from the maize they pinch from arable fields, chopping down ornamental cherry trees in public parks, punching their deep burrows into flood walls, undermining roads or stuffing the outflow pipes from our sewage farms full of septic waste is beyond what we know. Beavers perform these ancient behaviours in modern environments where watercourses are narrow or shallow because the 40-million-year-old circuit boards in their heads instruct them to do so. To ceaselessly engineer their surrounding landscapes to ensure they suit their purpose.

While coexisting with beavers is, for those living now, a novel experience, prehistoric people knew the beaver well. Archaeological studies demonstrate that early settlers in Britain preferentially selected beaver-generated environments for their abundance of fish, waterfowl, large herbivores and other prey. If they incorporated islands then all the better, as these features protected the hunters from becoming the hunted at a time when the big cats were still kings.[1] Crannog dwellers in Britain lived and built their dwellings on the top of former beaver lodges. They walked out into their wetlands to gather gnawed timbers, sharpened into ready-made posts, and utilised them for structures of their own.[2] Some of the earliest animal effigies ever discovered worldwide are of beavers, and images of the Eurasian beaver's North American relative, *Castor canadensis*, feature strongly in the mythology and legends of many indigenous peoples and their art.[3]

Beavers are the second largest species of rodent in the world. They are famous for their large front teeth, which they use to fell trees for dams, lodges and food. Although capable of dropping trees up to a meter in diameter, beavers will preferentially select much smaller material. They are an aquatic species, well adapted for a watery lifestyle. If well maintained, their near-impermeable coats can trap a bubble of air around them when they dive. Beavers are also vegetarian; they never eat fish, choosing instead to browse on a diet of grasses, forbs and other aquatic plants in the spring and summer before switching to the bark of the fine upper branches of trees or shrubs in the autumn and winter. They do not hibernate and must therefore work hard to form a cache of this material which they sink, stick and weave into the mud of the water body next to their abodes, before the advent of snow or ice prevents its gathering. They rely near completely on this established reserve for survival until the warming soils of springtime prompt fresh new growth.[4]

As I've witnessed many times, beavers are caring creatures. They love their babies. Beaver families defend and nurture dependent offspring. Although their mother's milk is only imperative for the first few weeks of life, kits are still dependent for at least their initial year on older siblings and parents. These aunts and uncles prevent them from swimming in water they consider to be hazardous and if they roam too far, carry them struggling, clasped in their front paws, back to their lodges. They cuddle them, groom them, whisper comfort in their soft, downy ears, curl up with them daily and essentially through the first winter warm them in their great snuggled huddle of a communal nest. They make beds for them, gather food for them, protect them from predators, afford a caring home, tolerate their tantrums.

When beaver kits die, as they sometimes do, there is even evidence that their mothers will on occasion try to bury their tiny cadavers if they can. We have no knowledge as to why they perform this behaviour, but recent footage from Switzerland demonstrates that they undertake this task with extreme care.[5] While it may not be prompted by 'love', it is a sentient act that is moving in the extreme.

If you're a 'Beaver Nut' and realise earnestly just how critical these creatures are to the future well-being of the earth, with a pivotal role in the creation of abundant biodiversity, water provision, purification, flood and drought alleviation, you will pursue beaver advocacy with the kind of tedious zeal generally restricted to deluded members of obscure religious cults. But no matter how obviously clear it is to you, understand that it is not that obvious to most other people. While your loved ones, parents, wider families and understanding friends may have to tolerate your views, many other people with lives filled with more absorbing interests such as stamp collecting or making tiny bedside lamps out of seashells will not. People have been killing beavers for so long now it's considered by most to be completely normal, commonplace and commercially appropriate.

It's no mystery why we killed beavers in the past; we know exactly why. Their value was considerable. Our insatiable lust for their glands, furs and meat drove their demonic destruction. Once native from Britain in the west to China in the east, from the upper rim of the Mediterranean in the south to the edge of the Arctic Circle in the north, beavers were hunted. Unremittingly. Without remorse. By the time of the Romans, their range was fractured. While some central European powers tried hard to protect their populations – for their commercial worth – by appointing court officials called *Beverari* to administer all matters pertinent to the beaver, most did nothing.[6] Their pyre of destruction burnt white-hot.

———

There are many folktales about beavers. It was once believed quite widely that when pursued a hunted beaver would castrate itself with its teeth to 'ransom his body by sacrifice of a part, he throws away that, which by natural instinct he knows to be the object sought for. . . . And if by chance the dogs should chase an animal which has been previously castrated, he has the sagacity to run to an elevated spot, and there lifting up his leg, shews the hunter that the object of his pursuit is gone.'[7]

Unfortunately for believers of this legend the substance sought by hunters, castoreum, is not derived from a beaver's testes but rather from its scent glands, which are located – one on either side – in the internal lining of their cloaca. Castoreum can contain a high concentration of salicylic acid, the main ingredient in aspirin, which is derived from willow bark. If the beavers concerned have been feeding quite commonly on this shrub, castoreum works well. If they have not, then it is less effective. Although the ancients did not understand this, they did know that castoreum's properties were variable, and may at least in part have attributed any variability to human forgery. The latin name for beaver, *castor*, is derived from the Greek *kastor*. Kastor was one of the divine twins (with Pollux being the other) who were worshiped by women in ancient times as a healer and preserver from disease.

While qastoriun, qasturiun qastur, quastura or more exceptionally 'jundubadastur' – drawn from the Persian *gond* for testicle and *badastar* for beaver – was recognised in the early Middle Ages for its medicinal utility, the farther away from real beavers the dried product was traded, the more confusion surrounded its origins.[8] While the Sephardic Jewish philosopher Moses ben Maimon (or Maimonides, 1138–1204) is believed to have seen a live beaver, Dawud al-Antaki (or David of Antioch, 1543–1599), a Syrian Christian physician, who described the beaver as 'a small, wild animal covered with black hair used for medicinal applications such as headaches and earaches, to treat diseases of the liver and spleen, leprosy and pus in the eyes', could not have done so as he was blind.[9]

The zoologist Al-Damiri (1341–1405) described his jundabadastur as having 'no forepaws, but he has hindlegs and a long tail. His head is like a human head. . . . He crawls on his chest. . . . ' While this outlandish feature mix of seals and otters was of course fantastical, Al-Damiri was exact in his description of the glands: 'He has four testicles; two outer/apparent ones and two inner/hidden ones. . . . Inside his testicle there is something like blood or honey.'[10] These after all were the important parts. The aspect of human interest. The bits that people paid for.

———

The castoreum trade, like that of any other valuable commodity, was international, and when combined with the desire for fur was the principal driver of the beaver's old-world destruction. In North-Central Greece, for example, the city of Kastoria ('place of beavers') was once famous for processing their pelts. Here the craft of fur preparation was developed to a virtual art form in Byzantine times under the sanctified icon of its protector, the prophet Elijah.[11] In the Greek language the name of the Kastorian Fur Association translates as 'O Prophetes Elias' after this luminary. While beavers are extinct in modern Greece and none live anywhere near Kastoria anymore, 'baby' beaver furs are still available for sale at the International Fur Fair, which is held annually in May, where they are modelled by thin, blonde, Russian women with darkened eyes and wide-winged brows. You can take your pick from a wide range of lurid colours and designs. In Britain, the change in the public perception of fur from a necessity to a luxury marked the beginning of the historic fur trade. In medieval times furs were considered so valuable that their use was strictly controlled by a series of 'sumptuary' laws enacted between 1300 and 1600. The London Skinners' Charter of 1438 brought in legislation to control the size of furs to be used, where and how they could be worn, and which types of fur might be used for edging and lining garments. For example, only high-ranking clergymen were allowed to wear any furs, including sable, beaver, marten and genet, with ermine being reserved for the nobility. While the middle classes were restricted to wearing furs of a lesser value, commoners were allowed only garments made from lambskins, conies or cats.[12]

During the reign of Henry II (1154–89), the workers in the fur and leather trade who were skilled in dressing skins and making articles from them were described as 'pelliparii' or 'peleters'. They formed the first Skinners' Guild for the wealthier merchants who bought stocks of raw skins, to dress them and to create products that could then be sold in their own shops. These aggregations of activities were often located in a particular area of a city such as Skinners' Row in modern Lincoln.[13] Like the beavers, the Skinners had their own saintly patrons such as St Petroc, whose stained-glass face presided over their works. Is it worth wondering if he ever met Felix? Would they have known the same beaver?

By the late Middle Ages, supplies of Eurasian beaver furs were becoming increasingly sparse and the only source remaining in seemingly inexhaustible abundance was in North America.

While the First Peoples of what are now Canada and the United States had always killed beavers when they wanted their meat, fur or body parts for a specific reason, they at no time sought to exterminate them as a resource. Commonly their hunters revered them more than any other animal and would hang their often-purposeless forelegs, after decorating them, in a tree to prevent their consumption by scavengers.[14]

The white colonists were different. Fur was money. A literal currency where 'made' pelts were interchangeable for goods from the buyers at the trading posts. They were bound in bales for transport to the auction houses of the east before beginning their journey to processing centres worldwide. The soft underfur or 'beaver wool' was the desirable part and its separation required ingenuity in a time before mechanisation provided its own complex commercial solutions. First Nations women were paid a pittance to sit outside the forts of the fur companies with the pelts layered over their laps to pluck out the guard hairs. Others were encouraged to make loose fitting garments with the skin-side out, which they wore over winter to allow the friction of their bodies to remove the stubborn material. Although greasy, rank and verminous, these plucked pelts were worth way more than the original pelts.[15]

The underfur was felted and made into hats in specialised manufacturing centres. Some, such as Denton near

Manchester, retain the image of a beaver on their town's coat of arms in a tribute to the good times gone. Beaver hats were durable and waterproof. They were utilised by merchant's guilds, the navy and the army. Everybody who was anybody wore a beaver felt hat.

They were the favourite hats of Charles I. In 1638, he incorporated the guild of 'bever-makers' and prohibited the importation of hats, in order to support the domestic felting industry. Prudently, John Bradshaw (1602–59), the judge who presided over the king's trial and execution, wore a 'broad brimmed, bullet proof beaver hat, which he had covered over with velvet and lined . . . with steel.' He also wore armour under his robes.[16]

Companies such as the Hudson's Bay Company, the North West Company or John Jacob Astor's American Fur Company competed for supreme control of this lucrative market. They were ruthless. When one felt threatened by the predatory intents of competitors in territories they considered their own, they would urge their trappers to overtrap. To abandon any febrile pretense of sustainability and literally, practically, kill every furbearing creature with the specific intent of creating a 'fur desert'. Once this aim was achieved and their warehouses were full to overflowing, they were faced with the inevitable dilemma of what to do with their haul. If released in totem it could depress the markets unsatisfactorily and therefore their solution was both simple and obvious – to burn them.[17]

The pain. The misery. The fear. The destruction and death this caused bothered them not at all.

When the fur played out, the trappers went, too. Debased and abandoned.

———

In addition to being hunted and trapped for castoreum and fur, beavers were also eaten because the Roman Church declared they were fish. The medieval historian Gerald of Wales stated 'that great and religious persons, in times of fasting, eat the tails of this fish-like animal, as having both the taste and colour of fish.'[18] Centuries later, English clergyman and author Edward Topsell recorded in *The History of the Four-Footed Beasts and Serpents* that beaver 'tails have weighed four pound weight, and they are accounted a very delicate dish, for being dressed they eat like barbles'. He goes on to describe that 'the manner of their dressing is, first roasting, and afterward seething in a open pot, that so the evil vapor may go away, and some pottage made with saffron, other with ginger, and many with brine; it is certain that the tail and forefeet taste very sweet'.[19] By the late seventeenth century it was no longer just the tail that was allowed on fast days but the whole beaver itself. When the Bishop of Quebec asked his superiors whether his parish could eat beavers on Fridays during Lent, the church declared that indeed they could for the 'beaver was a fish due to the fact that it was an excellent swimmer'.[20] And so the killing continued.

While Topsell did repeat ancient and fantastic legends in his writing, his account of the biology of beavers is accurate in parts. He describes in familiar fashion how they are not much bigger than a 'Countrey dog, their head short, their ears very small and round, their teeth very long,' how they used the same paths to and fro from the water, how they ate 'the bitter rindes of trees, which are unto them the most delicate, especially the Aldren, Poplar and Willow.' He also provided insight into how beavers were hunted, recording that '[beavers] are taken for their skins, tails and cods . . . when their calves (caves) are found there is made a great hole or breach therein, where into is put a little dog, which the beast espying, flyeth to the end of her den, and there defendeth herself by her teeth, till all her structure or building be rased, and she laid open to her enemies, who with such instruments as they have preset, beat her to death'. and that 'They cannot dive long time underwater but must put up their heads for breath, which being espied by them that beset them, they kill them with gunshot, or pierce them with Otters spears.'[21]

———

Despite the carnage, it is clear now that even in modern Britain we have never lived in an entirely beaver-free landscape. Long after they ceased to exist as living beings, their submerged skulls and wood workings continued to surface and surprise from the mirk of the old fen mires. An assembly of their skins, stitched together with their outsides turned in, cradled and caressed the delicate intricacy of the lyre in the Anglo-Saxon grave good assembly of the king buried in his ship at Sutton Hoo.[22]

Sometimes the landscape remembers.

Recently, as I sat in a municipal office to consider yet another fenced beaver trial in the north of England, a site manager showed us a remarkable image. His aerial slide unwittingly displayed a beaver-generated landscape – a wet valley swamp on a stream called the Barbrook. *Bar* is derived from old Saxon for beaver, and although its beavers are now long gone, their past existence remains etched in an intricacy of the patterns left by the water's wanderings as it slewed and shifted a millennia ago in an ineffectual effort to bypass their dams. The multiplicity of bell-shaped structures on the meandering stream system that developed and is retained in the modern, now sheep-shorn bare landscape is the legacy of their works.

According to the site manager, these structures still fill rapidly and completely with water when the brook is in spate, meaning beaver dams fulfill their natural, flow-slowing function long, long after the death of their creators.

Revelatory images are seldom this stark. Our ploughing, or infilling, or drainage in the open land has generally destroyed them quite utterly. Where they remain, as the flat silt plains that once coagulated behind their dams with perhaps a fresh stream cutting through in forests, the trees softly mantle their memory.

But although visions of the sort seen at Barbook are rare, sometimes, other surprises occur. In 1837, a nearly complete skeleton of a beaver was found in a hole in the bank of the river Stour near Keyneston Mill in Dorset. The discovering archaeologist recorded that 'the hole in which the bones lay did not appear to have any communication with the surface above. . . . Slight as is this evidence, I am inclined to think that the animal entered its home underwater.' He was quite correct. Beavers, when they have no need to create their famous stick-nest lodges, will simply excavate a burrow system in the friable soils of a riverbank from either at or immediately under the water level via a tunnel upwards into a series of snug, dry living chambers at the top. Although they may still be occasionally exposed by erosion in British riverbanks, these features must be becoming rarer. If no direct beaver evidence remains, their emptiness is equivocal and this Dorset discovery remains the only example to date of any definite physical association with its creator.

Beaver dams, lodges, distinctively gnawed sticks or trees, bone fragments and their great orange incisor teeth set in amulets have, however, been unearthed from many other locations. It is inevitable that as more archaeologists become aware of the former abundance of beavers in Britain, and as a result more familiar with their field signs, that other evidence of existence will appear. Perhaps one day their presence will be confirmed in the extensive lake-lands of Ireland where to date there is no proof of them ever existing. The wealth of material we now have and our understanding of beaver ecology elsewhere renders inconceivable the prospect that they were not at one time present in all our watercourses, slowing flows, filtering silts and building soils. Readying the land for us. To use. To farm.

———

The past importance and prevalence of beavers on Britain's landscapes is also evident in our place names. From Manchester to Ashford, from Leith to London, beaver place names abound. These houses, roads, streets, closes and primary schools are the last legacy of their great exploitation.

Sometimes the memory is idiosyncratic. The Beaver Inn in Appledore on the North Devon Coast has been in existence for at least 400 years, and contains beaver memorabilia everywhere; while it has no obvious modern connection with beavers, an endearingly realistic clay effigy looks down on customers from the roof ridge high above. It's a lovely, lively local pub, warm and inviting in its wide bay windows when the winds from the Atlantic blow the stinging spray east. They do good food and folk music.

In the remote western part of Scotland, a tradition was recorded amongst the highlanders in the 1770s of the *Losleathan* (los-loy-dan), or broad-tailed otter, being once abundant in the region of Lochaber.[23] Two entries in the 1848–52 ordnance survey – *Coire Tolldobhrain* ('the Hollow of the Beavershole') and *Alllt*

Coire Toll-dobhrain ('burn of the hollow of the Beavers hole') – may also recall their memory. *Dobhran* means 'dweller in a wet place' in modern Gaelic, and these place names are not recorded in more modern maps.[24] Their original meaning was derived from the Ordnance Survey officers' verification of enquiries of local individuals of prominence such as clergymen, farmers, doctors and other prominent citizens. These names were specifically recalled by Alexander McBeath of Shieldaig and the Rev. K. Macdonald of Applecross.[25] In more modern times it has been suggested that they were misrecorded and referred to common otters instead. There is no reason to consider this modern interpretation correct, and other evidence from elsewhere demonstrates that when beavers become uncommon a similar confusion arises. It is likely that the memories of the old people who spoke first are most likely to be true.

There is a *Beverkae* place name in Fife, and although no other place names record the presence of beavers in Scotland, there are trade and oral records. Hector Boece for example, who was the first principal of Aberdeen University, writing in his *Scotorum Historiae* of 1526 recorded beavers as being one of a wider range of species to be found in the region of Lochness. In his time they were present in such numbers that their furs could support a trade with German merchants.[26]

In Wales the beaver was called the *afangc*, and many place names associated with this word still exist. In the North Welsh valley of Nant Ffrancon ('Vale of the

Beaver'), one particular location is named the *sarn yr afangc* ('the beavers' dam').[27]

Gerald of Wales (1146–1223), also known as Cambrensis, was of mixed Norman and Welsh descent. In addition to being a historian, he was an archdeacon of Brecon and a royal clerk, travelling and writing widely on many subjects. In or around 1170, he was wandering around Britain with the then Archbishop of Canterbury trying to drum up at least some enthusiasm for another crusade when he encountered beavers. His description of the landscape they occupied is illuminating. 'The Church dedicated to St. Ludoc, mill, bridge, salmon leap, an orchard with a delightful garden, all stand together on a small plot of ground. The Teivi has another peculiarity, being the only river in Wales, or even in England, which has beavers, in Scotland they are said to be found in one river, but are very scarce.'[28]

It is quite clear that he was describing no wilderness.

As he went on to recount 'the manner in which they bring their materials to the water, and with what skill they connect them in the construction of their dwellings in the midst of rivers', it's obvious to wonder if the people who harvested the salmon sustainably were also trying to do so with the beavers.[29] Why else would they remain on the Teivi when they were increasingly absent elsewhere?

The many English place names and their derivatives from the fifth century onwards – *bar, bjorr, beofer, beuerlic, beuer, befer, bewer* – mask others that are not so obvious. On north Exmoor both a *bibers* and *bibors* hill were identified in the first series of Ordnance Survey maps, from 1805–68. Together with upper and lower Beverton and Beverton pool, all are located on tributaries of the main river Exe. Although both no longer exist, they may indicate the presence of beavers in times not so long lost. If *Beaverdyke* in North Yorkshire describes well the muddy impoundments they create from plant roots and vegetation when trees are scarce, *Beaverhole* must recall their burrows and *Beverbrook* their streams. Other recollections summon memory of wider ghost-lands. While *Bevercotes* in Nottinghamshire refers directly to the place where the beavers build their 'cotes or dewllings', could it be that Beverton on Exmoor reflects a first human impression from high, wooded valleys where medieval hunters looking down at the 'townships' made by the beavers below saw reflected in their semi-order an effigy of their own ramshackle abodes?[30]

Perhaps Beverston in Gloucestershire is a long-lost trading centre in the dried castor sacs, 'cods' or stones from the beavers that were once so abundant in the vastness of the nearby reed swamps? In sleepy Suffolk the current Little Glemham Mill was once called Beaversham Mill. Was it built on the site of a former dam?[31] Did the old Devon word *wirth*, taken by colonists to Labrador for the cache of sticks that they gathered outside their lodge entrances for winter food, really survive for a millennium without human recollection of their activity in a county where only a single archaeological record exists?[32] Again, we cannot tell but it is more likely than not that direct individual experience and intimate impression is reflected in at least some of these names. What is for certain is that when the beavers went, much else went with them.

———

In the early sixteenth century, the King's Antiquary John Leyland wrote that 'beavers used to abound in the waters of the River Hull.' Beverley on Humberside means beaver meadows or stream.[33] In the flat vastness of its surrounding landscape the rare names of Stork Hill and Stork Dyke provide a tantalising testament to the lost life of the wetlands, which desiccated and dried with the passing of the beaver.[34]

In prehistory male moose would have swayed through swamps where stands of straggling greater tussock sedge, old and close-packed, stood tall enough to turn their antlers.

Shoals of silver fish would have plumed and pulsated in gin-clear lagoons as they raced to avoid predators lurking deep in their depths.

Torpedo-long eel-pout.

Great pike of vast size.

Porpoise and seals well into inland.

Otters in gamboling families.

Bears shambling in season.

Wolves on willowed islets where the water allowed.

Emerald frogs in call cacophony suppressing all spring sound.

In sunlight long snakes wrapped around branches with pond turtles beneath them basking in log-laden clusters.

Bronze burnished large coppers. Swooping swallow-tails bejeweled.

Clouds of spinning insects in translucent tornados strobing slowly upwards towards the lazy light.

All subordinated to the birds.

The solemn sentinels of the black storks in the leaf dim gaps on the limbs of the high trees. The fractured yarning of the erne soaring high in the egg-blue thermals. The reeling, piping flocks of whirling waders. The rising multitudes of dabbling ducks. Grunting colonies of ivory spoonbills. Snapping pink pelicans with bucket-billed beaks. Grey flocks of geese in honking formation. Booming bitterns, bugling cranes, chittering finch flocks and sweet singing warblers. The spring bill clattering of the white storks in their tall tree nests, guano splattered and ancient would have together combined the melody of lost life-lands anthem.

But for a long time, beavers still lingered.

When in 1577 William Harrison the Canon of Windsor wrote in his Elizabethan description of mammalian vermin that 'I might here intreat largelie of other vermine . . . and likewise of the beuer', he presented an account of the beaver he knew with stunning lucidity: 'Certes the taile of this beast is like vnto a thin whetstone, as the bodie vnto a monsterous rat: the beast also it selfe is of such force in the teeth, that it will gnaw a hole through a thicke planke, or shere through a dubble billet in a night'.[35]

I have in my time moved, trapped and handled many beavers. For those of us who work with them and know their abilities well, Harrison's description – echoing back through 400 years of time – is eerily, evocatively, exact. It is unlikely that this was a secondhand account as the impression of their power he described is vivid. Unforgettable. His story is not repeated elsewhere and it is not irrational to assume that whatever he witnessed he never forgot.

Harrison went on to state that their 'said tailes are a delicate dish, and their stones of such medicinable force, that . . . foure men smelling vnto them each after did bleed at the nose through their attractive force. . . . There is greatest plentie of them in Persia, cheefelie about Balascham, from whence they and their dried cods are brought into all quarters of the world, though not without some forgerie by such as provide them.'

In 1566, the 'Acte for the preservation of Grayne' saw a system of bounty payments made by parish constables for the heads of specific wild animals and birds considered, as Harrison referenced, to be vermin. These records, which afford a horrific account of loss, also provide evidence that beavers survived in small numbers in some parts of Britain into near-modern times. At Bolton Percy, near York, a church warden's account of 1789 records the sum of twopence being paid to a John Swail for a 'bever head'.[36] On the following page an entry records a shilling being paid to another recipient for an otter. Bolton Percy is connected by the Rivers Wharf and

Washburn to sites near Harrogate where the place names *Beaver Hole* and *Beaver Dyke* remain. In 1904, another author, Edgar Bogg, recorded that he had been told by an elderly man that his grandfather recalled beavers living in a location called Oak Beck in his youth. Bogg dated this time to around 1750, which would have been well within the time range of the slightly later record from Bolton Percy.[37]

Intriguingly, while Bogg goes on to state that 'scientific authorities on the British fauna say that the reward (two or threepence) down to late times paid by wood-reeves and constables for each bever-head (as the parish records for many northern places spelt it) was properly for the flat-nosed otter', the parish clerk at Bolton Percy quite clearly knew the difference and paid accordingly.[38] Otters were worth more so why would you accept tuppence for an animal that was worth a shilling? The wood-reeves were forest administrators, not gamekeepers. Their remit was to protect forest income and as a result it is also much more likely at a time when 'withy' was a valuable commodity that 'bevers' would have concerned them. Otters as pescatarians would have troubled them not at all.

After this time there are no more records of beaver in Britain. While it cannot be considered that the above, however grainy, reflects the absolute end of the beaver, by the early 1800s the works of the great Dutch drainers such as Cornelius Vermuyden (1595–1677) were more than a century old, and the canalisation and drainage of the smaller wetlands was also well underway. When the romantic poet John Clare (1793–1864) recorded that 'they hung the moles for traitors – though the brook is running still it runs a naked brook, cold and chill', it is highly unlikely that, outside what scrub remained on the big river systems, much sanctuary of worth would have remained for beavers.[39] By then the dominion of humans was near-absolute, and the large mammals that survived such as the red fox or deer only did so at the behest of those who insisted they remain for the chase.

Where and when the last beaver died, we will never know. Perhaps sad and alone. The last of its lost colony without care or company. With limbs stiff from old wounds. No warmth from its fellows, no honied aromas or last soft murmurs as its heart slowly stilled.

Or perhaps brutal and sudden. Voices, shouting, the burrow roof giving way, cold air rushing in and with it the hot rank smell of its pursuers snarling, snapping curs.

Saliva and sweat. Swearing.

Raining, crushing blows. Numb paralysis.

Blood and darkness.

A terror-filled death.

CHAPTER TWO

Spurting Streams of Grease

G ROWING UP, I HAD NO KNOWLEDGE OF BEAVERS. I do have dim memories of trips from my parents' home in Dundee to Edinburgh Zoo to see penguin parades, and I can remember a gigantic elephant seal that spent its long, sad life basking on a slime-stained concrete shelf above a pea-green pool, as both animals provided facile entertainment. The first when marching in comedic ranks, reeking of rancid fish, through a screaming crowd of ecstatic children, and the second because when it farted and sneezed simultaneously, it fired, from its wibbling proboscis, streams of olive snot across its enclosure to land with a satisfying splat on the rock walls beyond!

There were elephants on their own in cold stone sheds. Giant apes lounged corpulent in their heated house. Tame hyenas, abandoned by their old colonial owners who retired to Morningside to die, paced gibbering against the steel bars of their cages.

Were there beavers there then? Apparently so. As my infectious enthusiasm for the species developed in later life, my mum recalled a small grassy pen with a stream flowing through where – although nothing was ever seen – a large sign confidently proclaimed it to be the home of a family of the Canadian sort. This exhibition obviously made an unmemorable impression on viewers in general, and its location has subsequently been replaced by a much more inviting shrubbery.

The baby book kept by my mum and filled to the gills with my first hair, some rather dull artwork and notes on my steady live weight gain also records that on reaching adulthood I wanted to be a zookeeper. Every weekend without fail I would drag my bored parents to the Camperdown Wildlife Centre near Dundee to gaze with wonder at 'Jeremy the Sugar Puff' bear. In his redundant years, disenchanted with his role as an ambassador for sticky breakfast cereals, Jeremy became distinctly grouchy and as a result lived in a small, secure brick pit with an attractive wrought metal frontage to provide him with a view of nothing much more than screaming kids. Day in, day out. Today. Tomorrow. For the rest of his long, long life.

Other highlights included a one-legged flamingo – its other one had snapped off in the frost – and a hand-reared (and therefore, psychotically dangerous) roebuck that was always more than willing to try to stab you through the fence.

I left school when I was 17. I wanted no more regimented learning, and as I could not visualise the only option of art college as ever returning a living of any sort, I started work as a trainee livestock auctioneer with a firm in Edinburgh.

I had worked for years as a boy in my local auction market in the town of Biggar in the green lands of the Scottish borders. At a time well before health and safety was ever considered, I was one of the small, smelly kids who ran baying with the dogs up and down its alleys on Saturday sale days after the flocks of wild-wooled sheep, which tumbled in their thousands from wooden sided cattle lorries into the market pens. The autumn was their time of harvest when the lambs born in spring were sold. Their great-grandmothers from the same high heather hills, whose useful lives were done, accompanied them alert and aware with hoar-grey faces and coloured paint marks on their flanks. While the better ones might survive a few more breeding years on lowland farms, their poorer brethren would be minced by Monday afternoon.

Blackface-Scotch lambs, with their wool dyed a variety of fashionable ochres to make them fill the eyes of potential buyers more pleasingly, mingled with the fast-fading, squat, dark-nosed creams of the Cheviots from the south country. Bred by their owners for a show-ring perfection that required the selection of their largest and squarest, single-born ram lambs, they had lost their way as a breed and were even then fading fast.

Huge, rabbit-lugged Border Leicester rams with red numbers painted neatly on their broad backs, loped like camels up through the alleys to stand on their own in single pens where they could be better admired by stud buyers. Brown-headed Suffolks, trim and tidy Texels, bald-headed bluefaces with crimped, tight coats. A few white-nosed Swaledales with serrated swirling horns. Sheep for all purposes. Fashion and fancy.

I helped ancient farmers as they held the gates of pens to sort their jumbled flocks into something resembling order before their time came to turn in the sale ring. It was hard work that began early as soon as the first light of the frosty mornings allowed. Sometimes they paid me single pounds or coppers, and on days when this happened in the lull before the sale, I would run across the road, as the market with its round ring of sweet-smelling sawdust was firmly in the town centre, to the baker's. When you opened the door with its tinkling bell you stepped straight into a world of delight. Pineapple upside-down cakes, chocolate bombs and coconut snowballs were stacked in serried ranks on paper doilies in clean glass cabinets. The warm scent of the nut-fresh bread wafted sweet from the back ovens of the bakery.

My target was the Scotch pies. Leaking and bubbling from the heat of their creation in circular pastry cases topped with onions, mashed potatoes or beans, they contained a highly spiced infill of blue-grey minced mutton. Bitten into without care they could spurt streams of grease. They were delicious with mugs of hot, sugared tea.

The time was full of characters: wizened shepherds who only left their hill hirsels once a year; old, shambling big men who in the Great War broke Irish chargers for the cavalry; colonels farming with military precision; returned rubber planters keeping pigs; market gardening Poles who never went home after Stalin stole their country. They were all there, and through them the old-style farmers.

Wearing sacks over their shoulders on wet days, and walking in boots shod with the same nails as the cart horses whose pace they still followed, they did nothing fast. Experience had taught them economy of energy, and they thought too long for the young to be bothered. We gave them nicknames and laughed at the lowness of their livestock and lives. When they died, a brasher generation supported by the financial certainties of the Common Agricultural Policy filled their space at the ringside.

A kaleidoscope of colour reduced rapidly to a palette of slate.

The auctioneer was the man who ran the show. If he was cute of mind, he was the spectacle the farmers came to play. Although ostensibly he was there to get you, the customer, the best price he could, he was seldom trusted as he needed the custom of the big farmers in the long term more than all else, and cared not much for the small. His cajoling was best at the monthly sales of furniture and bric-a-brac when the sums spent were so low that there was nothing much to lose. There the ringmaster became the clown, who exerted humour in an effort to tease out the last sums of pence.

Wm. Bosomworth and Sons was a family firm of auctioneers in Edinburgh run by the stern, bent figure of its Victorian boss, Cliff. Pomaded white hair and tight clipped moustache. An always smart suit covered by a brown warehouse coat on normal sale days. Clean white for pedigree parades. He was a fair man but hard. A canopy of great wrought iron pillars covered his cattle market. Low, once-sky-blue painted pens for sheep, taller enclosures for cattle. Holding sheds, sheep dips, loading ramps, shedders. All the paraphernalia required for processing animals from life swiftly into death. It was an underworld.

The business was located on a system of cobbled roads that laired round a sprawling abattoir in the reeking west end of the city. Railway slidings and sick sweet-smelling breweries. A tannery and abandoned corn-exchange. Diesel fumes. Working men's clubs where the poor drank black and tans. Worn-out strippers on a Friday night.

At the time I thought nothing of transit from my part-time joy as a child to a trainee livestock auctioneer and valuer. It was an unforgettable experience.

The long hard days of heavy work made brutes of the yard hands. They were drinkers whose beer and whiskey-nip breath in the mornings was nauseating. They could sort animals quickly into their sale lots for auction, but it was efficiency that drove their thinking, not compassion. When things went wrong as they frequently did, they would kick, punch and bludgeon. They could not kill as that cost money, but they could hit apprentices who did not do their rage-blind bidding swiftly, and they did.

Spurting Streams of Grease

I learnt very quickly, as the bosses turned away and the spectators who were there never saw, to stay well out of their way. They punched and kicked others more than they did me, but when you are small there is little you can do. I loathed life there. It dragged me down, and I left to work again in the market at Biggar after two years had passed. Bad dreams sleeping are when you go back.

Through all this time I kept rare breeds of sheep in small fields next to my mother's home. Gingerbread Shetlands with their small curling horns, multi-horned Hebrideans with Sicilian lemon eyes and dark chocolate Soay sheep with exquisite cream tummies and rump patches. Old breeds with histories that stretched long back into the cultural warp of Britain.

I clearly remember being in the market one day in Edinburgh when the only flock of old Lewis sheep I have ever seen were dispersed from their lunar island landscape. Gigantic rams with flowing fleeces and circular horns, elfin lambs with their wild-eyed mothers stepped straight out of a Victorian landscape painting and into the old railed ring of the auction market. Although some of the elders watching knew them from the far past, they did not care. Younger farmers laughed, and for a pittance they were bundled bleating out of time to their life's-end in a midland's abattoir.

I knew what they were. I knew the history of the highland clearances that their arrival began. There was nothing I could do. To plead with the farmers was pointless. I had no money and no ability to alter an outcome.

To stop it or say no. I have never forgotten that impotence. It still makes me sad.

I left the market and farming without any regret in 1988 when Dr Brian Thomson, a friend who was in charge of a country park on the outskirts of Cumbernauld in central Scotland, offered me the opportunity to manage its collection of rare breed domestic livestock. The job was simply wondrous. Never dreamt. As well as being able to care for herds of hairy belted Galloways and long-horned highland cattle, I was tasked with taking school and community groups out into the thousand acres of moor and woodland that surrounded the park's central buildings. I was there to explain how nature worked. I rediscovered the lost childhood joys of pond dipping for water scorpions and newts. There were field talks, open farm weekends, art and illustration opportunities. Interesting people with broad minds who encouraged you on.

The country park had a small zoo. Not a good one – a legacy of the original director, it maintained only native species. Tame foxes, mink, a few disabled owls, red deer and some old Russian wolves. Brian wanted to see this change. In 1990, he arranged for me to attend the international summer school on captive breeding endangered species at the Jersey Wildlife Preservation Trust – Gerald Durrell's zoo on Jersey.

There I sat in rooms with people from all over the world: serious Indians with round rimmed glasses; irreverent Australians who played stupidly hilarious practical jokes; big, brash Americans from Dallas; the discomfited director of India's Mysore Zoo, who as a follower of some entirely novel sect only ate the leaves of a specific Indian tree type. While he could without enthusiasm consume raw green vegetables, I can still recall his slight, trim person walking dressed in white into the sunset after our meal was done to try to identify in the Jersey hedges more palatable provender.

We talked about pink pigeons, diminished to a dozen by non-native crab-eating macaques on their home island of Mauritius. The story of how the last Mauritius kestrels, once reduced to a world population of 6, were recovered by the Trust from imminent extinction. Experts explained how they were attempting to restore the ecology of a ruined atoll called Round Island, where introduced rabbits and goats left to provide a source of fresh meat for sailing ships had expanded their populations unchecked. Their grazing had all but destroyed the island's delicate vegetation. Most of its endemic wildlife was extinct, but there on Jersey, breeding in Tupperware containers in a special isolation unit, were the last of its tiny boas and skinks.

It was Gerald Durrell's vision that one day they, along with the sad-eyed Alaotran gentle lemurs, the gilded golden lion tamarins, the aye-ayes with their skeletal probing central fingers and the humble Madagascan pochards, would all go back to where they once belonged.

One day when the world was a better place.

There were failures of course never mentioned in his books, such as the sad volcano rabbits that never did well. But there were many more successes. Beacons of the brightest hope in a world where he knew the lights were going out.

We never met, and I know now he was imperfect, but his enduring gift to the world through his actions and writings was the spirit of inspiration. It inspired me and many others.

Most of the course tutors were from England and one Dr Pat Morris, an erect, hawklike man with a wit that could wound, was the chairman of the Mammal Society. His concerns were focused on a range of native species that he firmly felt might benefit from Jersey's approach. He explained how, in his view, the conservation of many British mammal species was haphazard and immature. How much more could and should be done. My perception, if I had one, that somehow surely at home that all was sorted, thought through and sound, was disabused. It was obviously not.

In 1994, I was employed by the Sea Life Centre to develop their first short-lived British Wildlife Zoo in the New Forest. It was a complex undertaking, as many of the species they wished to display had never been maintained in captivity before, and little was known regarding their breeding or care. Moles, weasels, sand lizards, common dormice and water shrews all had their own fiddly, intricate requirements. As the imaginative underwater, underground or otherwise cleverly naturalistic exhibitions they wished to create were going to be purpose-built for these creatures, we had to get our abilities right.

Spurting Streams of Grease

One species that we wished to breed provoked a strong early memory. When I was small, I lived in Dundee. At around the age of 6 during the summer holidays, I was fishing for minnows with my brother Douglas in a stream next to a caravan park in Coupar Angus. Absorbed as we were in our hunt, we did not notice, until they fell shrieking from a banking above, a battling pair of water voles. They landed at our feet and separated. We dropped our jam jars and nets and fled in floods of tears back to our dad's rusty, white caravanette. Although I know now it was probably a determined male trying to knock his pugnacious mate into the water for some swift, safe vole sex, I did not know that then. It was an image of horror I never forgot.

Water voles were a species, which, in 1994, had seldom been captive bred. Although they were commonly considered to be abundant, Pat was one of the first to question the truth of this assumption. Habitat loss on a vast scale through the drainage of wetlands, the overgrazing down to the water's edge of riverbanks by domestic livestock and the growth unchecked of light-blocking riparian woodlands all conspired to remove the dense summer jungle of reeds, iris and other aquatic plants that both fed and protected water voles from their predators. Wherever we asked there were none to be had. We spent near a year knowing little and chasing shadows. Wherever we looked they had gone.

Their old burrows were everywhere. The hewn-out shapes of their runs and feeding platforms. All the evidence of long-lost occupancy. Derelict, overgrown, decayed. It was so bloody sad.

Although introduced North American mink were called culprits – and they played their part – the fault was all ours. We have created and continue to create in Britain through our own actions a landscape so utterly devoid of opportunity, so stripped of life resource that even the tiny creatures can no longer exist. We have tried to take everything. Pointless farmland created with expensive taxpayers' support to produce nothing except poverty now for some of its owners. We have ripped out ancient woodlands. Ploughed-under sites revered by generations of ancients until ours arrived. We have canalised rivers and reinforced their banks. Built isolating highways with killing concrete-central barriers. Poisoned the soils.

Destroyed unforgivably so much.

Time and again you see this truth.

The last of the field crickets, the tadpole shrimps, the black-dotted ladybird spiders, the shrikes, the ouzels. The flickering glow worms at night. The ululation of the new forest cicada.

The lights in Britain are fading fast. Faster than we can imagine.

———

Eventually we found our water voles in a fish farm in Hampshire. So many indeed that they were undermining its access roads and collapsing the soft soil banks of its fish ponds. The owner was happy to let us trap there and glad to see them go. It saved him the trouble of shooting them when he could.

We brought them back to our breeding facility and for six long years made many mistakes. The pens were too big and too natural. Once released into them we could not harvest effectively; what we discovered was a savagely territorial species. Only a single male and a single female could be put together in a large landscaped enclosure with banks to burrow into and swimming ponds. Any other combination and they fought to the death. Although they live wild in loose colonies, the adult females are horrors and neither the males nor their own offspring wanted much to do with their mothers as they matured. Left to their own devices unless the babies in their sequential litters – female water voles can have up to five litters in a single season – were removed and held separately elsewhere, their own older relatives, unchecked but limited by enclosure walls, would destroy them in the autumn.

We tried smaller pens. They escaped. We adjusted their depths and rats dug in. We did not realise their lifetimes were as short as they were, with females only surviving for a single season while males seldom exceeded the age of 2. We tried to keep individuals and bred them for too long. When they were worn out naturally and their life course run. This is all normal when you are learning novel skills. There is no shame in failure providing you don't make the same mistakes twice.

We improved and eventually, by 2001, had developed a system so reliable that we could breed water voles in numbers sufficient for reintroduction. The first release we tried was into the grounds of the Wildfowl and Wetlands Trust Centre at Barn Elms in London. The released individuals prospered and did well. One of their biological functions as a species of medium-size rodent is to provide an abundant prey base for a whole host of predators that rely on them for food. A large male water vole can weigh up to 330g (12 oz) while the next equivalent species, the field vole, even when mature, seldom exceeds 30g (1 oz).

As we watched with mounting anxiety, the grey herons speared our reintroduced wards with ease. We did not understand that this activity equalled success. One of the biggest problems wildlife now faces in environments entirely manipulated by us, where it has become the norm for silage to be cut from March through to November, is space to live. Even if you are a tiny harvest mouse and there are hundreds or even thousands of acres of grassland in the landscape surrounding your hedgebank burrow, your children can't expand out into this because when they do, the swift shaving blades of the field mowers obliterate them. If you avoid this, both your hedgerow and the unwanted strip of roadside grass you might otherwise utilise will be flailed bare with swivel-headed cutters in the autumn to ensure that no contour, no drain, no decaying stump or other small refuge remains. Your options diminish steadily with nearly every one of our normal actions. Even though we do not desire what we destroy, we will deny it to you.

From a predator's perspective, the despised brown rats are full of rodenticides. Eat too many when nothing else is left and their toxins will kill you, too. The rabbits have viral haemorrhagic fever and are fast fading, and the water voles once so abundant and so familiar that author Kenneth Grahame employed their well-worn image for his character of Ratty in *The Wind in the Willows* . . . well they, I am afraid, are swiftly approaching functional extinction. In 2004, there were estimated to be only 1.2 million left nationally. After an estimated decline of nearly 97 percent, now there may be less than 130,000.[1]

Durrell would have been appalled. Sickened and saddened that in a resource-rich nation such as ours we could not summon the will to save the water rat.

Water voles are important not just as a prey base. Without their presence in riverbanks, burrowing and excavating, mowing tidy vole lawns and felling taller plants, gnawing open seed pods and dispersing tiny rootlets at the water's edge, these environments are immensely poorer places by far. Their going will mean no more hiding places for grass snakes, no hibernacula for newts, no nesting spaces for kingfishers when their intricate mining activities no longer collapse banks to expose safe, steep tunneling sites.

In the end their going will also imperil the future, for if we ever get beavers right in Britain and this life-giving rodent returns on a much larger scale, it's the Tic Tac–size droppings in nutrient-rich vole latrines that enable the essential mycorrhizal fungi to reform in the sediments of their abandoned dams.[2] They are important. Very important. We can't afford to let them go. And yet that's exactly what we are doing at an astounding rate. It is no good.

In my time working with the species, I have captive bred more than 25,000. But I am losing. While some of the projects to which we have supplied water voles have failed, many more have succeeded beyond the wildest expectations of those of us who began their reintroduction. Some have created populations that are vast. Other experts have identified how to control mink strategically or restore lost habitats that often, quite simply, can be achieved by fencing out destructive livestock grazing from riverbanks. We know that beavers would help to restore the wetlands that so many other species so desperately require. We know that water voles are important beyond nostalgia.

When you are young and naïve you think the professionals involved will sit down soberly, look at the evidence wisely and then labour long to achieve a good end based on sound science and experience. As you get older you realise that it is commonly not this way. Reintroduction is now the only hope for the water vole. They are too far gone from most of their ancient haunts even if their decline could be halted where they still remain, and it can't. There is too much of our infrastructure in the way. Too many obstacles. There is not enough time, and left to themselves they will simply continue to dwindle and decline. Restoring them will not be easy but there is no plan B.

Yet there are those who say no. That we should not meddle. That it's stressful. That we must let nature take its course. That they should pass. There is nothing natural at all about the decline and demise of this species and it's not their patrician choice to make.

It's yours.

I have spent enough of my lifetime surveying the ruins of their landscapes. I have seen huge sums of money squandered. I have watched petty individuals in positions of power, who could have helped had they wished to do so, preen and prance. They are fools. In fools' costumes seeking respect alone for position. Their bell-bedecked headdresses ring forth quite clearly the emptiness of their minds. As technicians they have responsibility for the duties they fail to discharge. They measure their own progress through ensuring inaction. I have witnessed their stupidities. Protected by unionised employment, they retreat like mauling morays back into a complex reef of incompetence that affords them the refuge they require.

Afraid of their power to influence, refuse licences or cause critical delay, most of us say nothing. This affords succour, which we should seek to deny. Time, circumstance and society are on our side. Better to try than not. If we don't then they will just continue to feed fuel into the furnace of the train that takes both our natural world and all of us together hurtling over the precipice of destruction and into the depths of the dark.

On Jersey in 1990 I met the director of an Indonesian project set up by international funders to reintroduce a white mynah bird with a striking blue face – the Bali starling – back into the wild. The species had been so collected from its forest home as a desirable cage bird for its power of speech that by the early 1990s none remained in the forests anymore. Zoos and conservation organisations from all over the world had provided the director with breeding birds. One night when he was asleep in his home with his wife and children, a military coup took place. The first he knew of this was when an armoured car crashed straight through the gates of his living compound. He got out of bed and walked towards his front door, but before he got there, it was kicked open. A flashlight shone straight in his face blinding him, and as the clicks of the safety catches on the Kalashnikovs told him they were ready to use, he heard one of the gunmen say 'give us the birds'. Visibly trembling as he told us his tale, he said, 'I had no choice. I said allow me a moment and I will get the boxes.'

He never saw the starlings again.

We face nothing like this in Britain, although it is a common challenge for brave people fighting to conserve just a little tiny piece of nature's wonder elsewhere. On occasion they have died for their beliefs. On behalf of the elephants, rhinos, gorillas or butterflies.

I have found it's best to pursue your own route in life. To make up your own mind. To think for yourself. Through its course you will find friends of all sorts who will aid and assist you, sometimes at cost to themselves. Remember always in the petty world of pomposity that surrounds us that we will never have to face what these others have had to, and to try to do your very best.

Have no regard for the words of the witless.

They neither love nor care.

CHAPTER THREE

Popielno

T HE IDEA OF BRINGING BEAVERS BACK TO BRITAIN in the mid-1990s was not new. Various estates imported and maintained beavers as oddities at different times.

In the 1600s a single beaver in the menagerie of the tower of London was described by one of its onlookers as the most curious beast he had ever seen.

In the 1860s a Canadian beaver family was released into Sotterley Park in Suffolk and later destroyed when their constructions were considered to be an eyesore. When the survivors migrated from their release location into the nearby Benacre Broad, they, too, were killed, and of the last of the colony it was written that 'two were sent to London to be stuffed for Lady Gooch and the head keeper took the skin of the third.' By the mid-1870s they were all gone.[1]

In the mid-1860s an attempt may have been made to start a beaver farm, although the evidence at

present for this is not clear. If true, it is unlikely
that it was a philanthropic undertaking and was
probably a fur-based enterprise.

In 1874 it was the Marquess of Bute's turn. After
a failed breeding experiment between a pair
of French and a pair of Canadian beavers (the
Eurasian beaver is a separate species from the
Canadian with a different chromosome number
and as a result, even if they do breed in captivity,
their matings produce no offspring) that fought
savagely and did not survive long, a further group
of Canadians were acquired, which flourished for
a time.[2] These were enclosed in a large, barred
enclosure, which they rapidly denuded of trees.
Although provided with fresh fodder on a regular
basis, the colony eventually dwindled, with the
last dying in approximately 1890. The place name
Beaver wood, near Rothesay on the Isle of Bute,
commemorates their presence to this day.

Another colony established in a large enclosure in the
late 1800s by Sir E.G. Loder on his Leonardslee
Estate near Horsham survived until 1948. His
grandson could still remember them in 2014 quite
clearly as the gardeners grumbled constantly about
their requirement for a regular supply of trees. To the
best of his recollection the last of these beavers was
delivered to London Zoo not long after World War II.[3]

In 1946 a male Canadian beaver dubbed 'Jock' who
was gifted to the zoo by the Canadian Government

escaped into Grand Union Canal and was shot over 13 miles away by a Mr Fred Neighbour who assumed it was an otter and therefore worthy of instant death. Interestingly locals who visited its decomposing carcass before its origins were known expressed no surprise at its presence, and considered that it had simply been swept away from its home in the bank of a northern river by recent floods.

Lord Onslow, writing in the *Countryman* magazine in 1939 was the first to suggest that as the 'beaver have become extinct in England only within the last few centuries . . . there seems no reason at all why they should not be reintroduced.'[4] The sixth earl, William Onslow was an interesting figure who had a wide-ranging interest in the preservation of fauna, and although unheeded at the time his recommendation was perhaps the first to suggest that the species should be returned for nature conservation reasons to restore the depleted fauna of a proposed system of British national parks.

In the late 1960s, a near-remarkable project almost occurred. Bill Grant, who was a visionary district manager for the Forestry Commission (FC) at that time, was awarded a Winston Churchill Fellowship to study the recreational use of the North American national parks. At a time when the Commission's approach to the creation of its forest estate in Britain involved nothing much more cerebral than the blanket coverage of whichever cheap landscapes it could acquire with single species plantations of North American conifers, he returned with some very different ideas about forest use. In Grizedale Forest in the Lake District he established a theatre where the leading thespians of his time performed in forest settings. He commissioned and created sculpture trails and was very interested in reintroducing lost species of wildlife.

One of these was the beaver.

In the early 1960s, in collaboration with his forest conservator, Jack Chard, he devised a plan. Grizedale Forest was and still is a good choice of release site for beavers. It constitutes a valley with a river and agricultural land, much of which is owned by the FC. The lower slopes of both sides of the valley have broad leaved trees with planted conifers on the upper hills. Jack believed that any beavers, once introduced, were unlikely to stray over the ridges that contained the valley. To enhance the site for beavers, they planted willow round its tarns.

In late 1969, Jack's proposal for the introduction of beavers was submitted to the Nature Conservancy Council (NCC), the then government's nature conservation advisors. They took little time to consider and reject his idea, and G.G. Stewart, the commissioner responsible for the region, stated later in a letter that 'I wonder how much thought was really given to the matter. But perhaps I am using the wisdom of hindsight. It was with considerable sadness that I had to tell Jack Chard that he could not go ahead with his plans.'[5]

There are few people now left alive who recall this project, and the FC holds no records beyond a pencilled scribble on a sheet of paper in Grizedale's dusty archive, which states the project was abandoned due to quarantine issues. It would seem initially that the beavers were sought from Canada, but this changed when it was realised that these were the wrong type and an arrangement was made instead to import beavers from Sweden. Martin Noble, the former head keeper of the New Forest, was tasked with the removal of the quarantine pens erected to hold them when he started his career with the Commission in 1973. He did not do a very good job; in 2010, when the site was again assessed for beavers, the remains of the original enclosures were still quite clearly visible around the high tarn on a hilltop at Peat Moss.[6]

In 1977, *British Wildlife* magazine advanced a case for the restoration of beavers to celebrate the Silver Jubilee of Queen Elizabeth II. It is unlikely that she was ever aware of this gesture, and given the somewhat desultory record of royal interest in species reintroductions, it's pretty unlikely even now that it would have met with much enthusiasm – unless of course its chasing with hounds was possible. The wildlife artist Barry Driscoll was its principal proponent. His belief was that as beavers by then had been so widely restored in other European countries their return would be acceptable in Britain.[7] The former Lord Mayor of London and conservationist Sir Christopher Lever perplexingly put down this effort with some pride.[8]

In the early 1990s, the veteran tree expert Ted Green led a beaver trip to Switzerland. Now as old and gnarled as an Ent with a complex system of underpants containing so many forms of ancient fungi that the government's nature authorities are preparing to declare them Sites of Special Scientific Interest (SSSI), Ted was and is a pioneer. He understood with clarity the suite of relationships that riparian tree species had developed with the beaver, wrote articles for dull forest magazines about how we should learn to coppice like them and was determined to do what he could as an individual to promote their return.

His Swiss trip, which was stuffed to the brim with beaver sceptics, foundered however, when they identified to their great glee an ancient black poplar over a meter in girth that a beaver colony had happily collaborated to bring down. Despite Ted's correct assurance that beavers generally feed on shoots of much smaller diameters and that veteran trees of this sort could be protected quite effectively with wire mesh or rubberised anti-game paint with grit mixed through, his gloating audience gibbered with glee and focused on this single image alone.

The time, yet again, was not right.

But the dream refused to die.

———

The idea of reintroducing beavers to Britain, which was re-examined in 1994 by SNH as mentioned in the prologue, was in part the result of the British Government's decision to ratify the Berne Convention for the

Conservation of European Wildlife and Natural Habitats treaty. This legally binding international agreement obliged its participants to at least consider the restoration of extinct flora and fauna. While England and Wales shirked this intent with foot-dragging futility, in Scotland at least SNH decided to explore. Earnest studies were commissioned on the wolf, wild boar and beaver. While all concluded in the end that restoration in theory at least was practicable, the snarl of sheep-farming interests swiftly consigned the wolf study to the depths of the deepest vault the organisation possessed, where it languishes still to this day, cobwebbed in its crypt.

Wild boar were also considered impractical as the limited deciduous woodland cover in Scotland would not, it was felt, provide a viable population with adequate cover or an effective supply of tree seeds. Many years later, when numerous boar began to escape from specialist farms, they proved their emphatic disregard for this prediction by feasting on farmed crops and establishing populations in coniferous forests, which are now well distributed throughout Scotland and to a lesser extent in southern England as well. Although reviled for their success by organisations such as the National Pig Breeders' Association – some of whose members were probably in part responsible for their initial presence – nature conservation bodies have been slow to support their return. In Scotland, for example, official communiques insist, despite the genetic evidence, that they are no more than feral pigs pretty much on the basis that they did not fill in any necessary permits to exist. There is little doubt that one day this argument may well be used to detoxify the prospect, as swine fever sweeps Europe, of free-living boar being the first indigenous species we might attempt to eliminate twice.

Only beavers remained.

The nature conservation agencies in England and Wales – English Nature (EN) and the Countryside Commission for Wales (CCW) as they were then – eventually emerged timid and fawnlike from their cultural undergrowth to also express an interest. Surely the modest and humble beaver was a credible prospect. Its behavioural activities offered much to the landscape and as an uncontentious candidate its return would prove to be an easy win?

Meetings were organised to discuss, and a visit was made to examine the beaver habitats created in part from a small population that was released at random in the Parc naturel régional d'Armorique in Brittany in the 1960s by a priest who kept them as pets in his garden. Dr Grahame Bathe, a sober scientist from EN, was a reluctant attendee. While initially he really could not see the point in Brittany, in a landscape similar to Cornwall with small fields and ancient medieval hedgebanks, he had an epiphany that was not entirely Calvados inspired.

Bathe saw how the wet meadows the beavers created provided ideal habitats for rare marsh fritillary butterflies to weave the protective silk webs their caterpillars require on the flowering plant known as devil's-bit scabious. He saw the life changes, the frogs, the birds, the water voles nesting in grass clump islands whose footings were sub-merged by the shallow extending waters of the beaver dams. He marvelled at the re-formation of reed-bed and wetland habitats, long drained by people, by the counter activity of beavers.

He understood.

Bathe became an advocate and was supported for a time by his managers. He was marginalised when they lost interest, but by then it was just too late. For years after when he travelled in Europe on family holidays, he clandestinely researched the location of beavers and took his family to see the marvels they wrought. Eventually, on behalf of his sulking children who were bored quite thoroughly with this obsession, his wife forbade any fur-ther activity of this sort. While he continued for a time as his career developed to speak up for his enthusiasm, his words were in vain. As he walked into meetings, he would commonly hear the word *beaver* followed by long bouts of sniggering. Eventually he gave up.

Others did not. A chance conversation with Roy Dennis, who had come to hear of my interest in this issue, provided the option of a different course. Roy, an internationally respected nature conservationist who has achieved in a lifetime more than any other individual I know, is the force that protected the ospreys when their return to Loch Garten in the 1960s to breed was imper-illed by egg collectors. He extended and is still expanding their British range through reintroductions into southern England at Poole Harbour as I write.

He reintroduced the great sea eagle, the vulture 'with the sun in its eye', to its lost landscape of storm-bound cliff heads in the mists of the far north. He and his able field assistant Dr Tim Mackrill are now returning them to their oldest eyries on the under-cliffs of the Isle of Wight. Amongst many other achievements, he has restored red kites and advised those of us now working on the return of the white stork. When you consider his personal example of perseverance, it's as near a pure record of great good as any individual can produce. I feel hugely privileged to have met him, known him, worked with him and called him my friend.

Roy Dennis said do it. Find out where the beavers might come from, sort out the issues involved and just do it. Just make a start. So, I did.

———

In the mid-1990s there were no Eurasian beavers in British zoos – Canadians were commonly held as they were believed to afford better if still somewhat tedious exhibitions – and as far as was known at that time, only a single breeding group with no spare offspring was maintained at the Planckendael Wild Animal park in Belgium.

Even if beavers could be found, they would be required to be quarantined for six months for rabies in escape-proof enclosures in buildings where any access to other wildlife was not possible. Additional health checks would also be necessary, and while this process seemed straightforward, little was known regarding what care they would require during this time. Would these highly active creatures simply wilt and die? Or could we create conditions that were adequate enough to ensure their survival?

Although some answers were unclear, a raft of well-aired hypotheticals that made it seem so very difficult were, on closer examination, nonsense.

After consultation with old zookeepers and animal dealers, whose memories stretched back far enough in time to remember the black beavers coming from the Hudson's Bay Company as part of their charter's commitment to Queen Elizabeth II to supply her with two every time she set foot in Canada – another explanation perhaps for royal disdain especially when their last two decedents sent to the Gateshead Garden Festival in 1990 for the Beaver Scouts display were promptly named Adolf and Eva by their gleeful new owners – it began to sound simple.[9] I was sure it could be done.

A chance encounter with a Dutch conservationist at a conference in Hannover suggested that the beavers we were looking for could be found in the east, and in 1995, six years after the fall of the Berlin Wall, I went to Poland.

Near a tiny village called Popielno, in Poland's north-eastern Masurian Lakes region, a beaver farm had been built by the Polish Academy of Sciences after World War II to resupply them – initially as a furbearing asset and in more modern times for their ecological value. Could beavers be obtained from there? While it looked possible as far as the authorities in England were concerned, a clear understanding of what would be required at the Polish end was going to be essential.

At a time when travel and communications with Poland are now so easy, it is hard to remember that at that time they were not. Without the assistance of the director of Poland's Poznan Zoo, it would have been near impossible to organise. Faxes disappeared into thin air. Sometimes the phone lines to Popielno would work and you would get a perfect direct line to the academie's genial and gloriously moustached boss Zygmunt Gizejewski. Sometimes you would get a group of giggling children who would tell you that Zygmunt was dead.

Masuria was breathtaking. As a result of forty years of communism, very little had changed. While the Russians had imposed giant collective farms with cathedral-size cattle courts and limitless croplands, their former workers were abandoning these. Echoing and empty as the social order changed, nature was beginning to reassert itself, and where western agribusiness had not reached rapacious roots, its glories were spectacular.

As our train sped north towards the lake-lands, clouds of white storks danced like ballerinas with their carmine legs dangling behind Sulwaki horses with flaxen manes turning hay. Like a clearly seen wildcat, the storks hunted and impaled on their stabbing, sabre beaks the homeless common voles running from the ruins of their newly shorn cities.

Near-invisible corncrakes crexed in the scrub-bound, stubbly savannahs that bordered the small lane leading from the tiny station down to the academie's field centre. Golden orioles shrieked from tall poplars, red-backed shrikes fluttered furiously to grab and impale dragonflies or whirring cockchafers in the late evening sun. The short dusty walk was brim-full of nature. Glow worms twinkled when the light declined. The field centre itself was surrounded by forest. As I hiked towards the lights of its central block, a moon-blue herd of Konik horses maintained for trekking tourists whisked the last of the biting insects from their flanks with sooty tails. The zebra stripes on their back legs displayed their ancient ancestry as the descendants of the wild forest horse, the tarpan, which was hunted to extinction in the late 1900s.

Popielno was odd. It contained a diaspora of random biological experiments that had been designed and abandoned for reasons that were never entirely clear. In a conifer enclosure lived a gigantic wisent / North American bison / cattle hybrid. Black and surly with long horns, it existed in complete solitude. No one was sure about how or for what reason this complicated creature had been created.

Its purpose was obscure.

Przewalski's cross horses shared paddocks with exotic stags whose great branching antlers reached tall to the sky. Most of these had been developed as the byproducts of an energetic programme of hybridisation between the Asiatic wapiti or maral and any other species of female deer they could obtain that would stand still for long enough. They were a strange and deeply sinister bunch.

Tamed to enable handling so that their semen could be gathered, they were as a result utterly unafraid of people. The project's goal was to develop through experiment Schwarzeneggers of stags, which would afford abnormally abundant rolls of soft, dark velvet for traditional Chinese medicinals. The fact that their growing antlers had to be bloodily severed from their heads with a hacksaw in a crush to obtain this material did not seem to trouble anyone much.

If you have ever worked with captive deer, you will realise just how dangerous tame males can be, and these chaps were simply terrifying. Fast as lightning, muscular and gigantic, they spent most of their waking hours sharpening their antlers with a relish that was demonic. I have not enjoyed a walk through a parkland less.

While Zygmunt waxed lyrical on the qualities of ear-tagged individuals, they rolled their eyes with amazing indifference, lolled tongues from their mouths and gave every indication that their infliction upon us of a horrifyingly violent, screaming death was imminent. I moved swiftly towards the exit gate when the tour drew to conclusion on the basis that I needed to get to a toilet fast. Anything to get out and ingest the sweet air of survival on the other side of the fence.

The beaver building was vast. An aircraft hangar of asbestos and grim external, concrete enclosures, it was soviet and severe. When the door opened, the inside service passage along its right-hand side stretched off into dim infinity. You could not see its end.

It took some moments for my eyes to accustom to the dim. To the left the sound of lake water, crashing from gigantic rusting pipes into a series of deep swimming tanks, made conversation of any sort with their keeper Hubert Nieglowski all but impossible. Doe-eyed, red-haired and gangly, as he spoke no English and very little German, it was difficult for us to conduct any kind of conversation. Without the excellent interpretive skills of a veterinary student-in-residence, any understanding I might have gained of what was happening there would have been poor. As we entered, she explained the husbandry of the many beaver families this building contained in great detail. She spoke of the beavers it had provided in its operational heyday and just how well established the species now was in Poland.

When she had finished, Hubert took a torch from his pocket and gestured towards a compact concrete bunker roofed securely with large flagstones. When he slid one off to the side and shone his torch onto its contents, he illuminated a scene of pure pleasure. There on a bed of finely split wood, tucked tightly together was the most splendidly snug beaver family. Some slept twitching and snoring gently on their backs with tiny ones curled tightly into the folds of their sides like furry grapefruit, round and soft. Scented and warm they mewed and slowly began to unfurl as Hubert spoke gently and stroked them with the palm of his hand. Others slumbered until the movements of their bedfellows stirred them awake. There were black beavers – the first I had ever seen – and tiny gingerbread juveniles.

A tubby teenage 1-year-old turned upright to stretch itself flat, yawning with its forepaws extended and its incisors exposed. A baby whimpered and cried. Their large dark mother uncurled, and by doing so exposed the four pink nipples on her chest to be still engorged like chipolatas.

She spoke softly to Hubert and he spoke to her. Then he leant forward and lifted her carefully into his arms. The interpreter told me to sit outside on the step of the shed and when I did, he passed her to me. She settled into my lap with her great leather tail turned beneath and turned her head up towards my face looking down. I have seldom known such rapture.

She was fragrant of wood and herbs. Of the indescribable sweet smell of beavers that, once recognised, you can sometimes catch in a willow wood on a warm day in landscapes where you know – frustratingly – they are not. But should be.

She was love and warmth. She was mother and maternity. She was kind.

In the years since we met, through the practicality that has had to drive my actions and thoughts with regard to this species on occasion, I have never forgotten her. Just thinking about that moment now, I pause and remember that in my life that followed nothing was ever the same again.

I remember her. She was wonderful. She touched my heart.

We took her into the outside yard for photographs and she followed Hubert around calling gently. Talking to him in the same language she used for her kits. Eventually, when she tired and had eaten the fruit treats he provided, she turned back towards the portal where the water lapped inside and slid back into the depths to return to her family.

In the days that followed, we discussed – as we sought the dusks for moose, followed radio-collared wolf packs into their swampland lairs and waded bare legged into the memorably mosquito-laden pools to gaze at the gigantic lodges the beavers had engineered – the possibility of obtaining breeding beavers for maintenance in wildlife centres to begin with and free release in the end. We agreed it could be done and that Hubert would select families that autumn for us to import.

Our beaver beginning had begun.

D.J.Gow
APRIL 2020

— CHAPTER FOUR —

It's a Fat Otter

T HEY WERE THE FIRST BEAVERS I HAD EVER IMPORTED
and I was worried.

Although everything that could have been done
was done – including the construction of large, solid,
tin-lined crates to meet the strict legislation required for
international air animal transports – moving wild animals
is never a straightforward prospect, and when things
go wrong, they tend to do so fast. The very last thing I
wanted to do was to open a crate full of dead beavers.

Britain is unusual in that importing beavers requires
a period in which they are quarantined for rabies. This
means that they cannot simply be collected from a location in Europe, brought to Britain and then released.
Quarantines are not the norm in continental Europe,
where most reintroductions do not involve lengthy
delays between the beavers being captured and their
release elsewhere with perfunctory veterinary checks.

There are only two methods of getting beavers to Britain. The first is fast, simple and expensive, and involves

flying them from their country of origin to Heathrow, where there is an international animal quarantine centre capable of caring for all manner of animal life from guppies to gorillas. The second involves driving to Europe to collect them. In recent years most of the imported beavers have come from a man named Gerhard Schwab in Bavaria (more on him later) who has many surplus each year as a result of his organised management system that removes them from locations where their activities are intolerable. Experience gained over years has taught us that although it may seem heartless not to bring in old, injured or orphaned individuals there is simply no point. We want beavers that can survive quarantine and go on to live long and productive lives in Britain. The entire process is expensive and transporting poor individuals is a waste of resource. Stated simply – when selecting beavers, it's the fat and shiny ones that we take while the others are culled.

From Bavaria it's a ten-hour journey back to Calais. If you have several drivers this journey is not so bad, but traffic accidents, football matches or a variety of other vagaries can cause significant delays. Once we got lost in the red-light district of Düsseldorf where, despite a broad range of 'fruity' options available to visiting humans, we could find no source of sweet apples for our beavers.

On occasion, even though you have all the correct paperwork, the port authorities can produce surprises of their own. Sometimes they just want to show a beaver to their mates and then with airy French elegance simply wave you through without further ado. At other occasions they will glumly peruse the paperwork you have given them and insist that no boarding is possible unless you can confirm the validity of your details at 2.37 a.m. on a Sunday morning with the necessary officials in Britain. This, in case you are wondering, is a simply impossible request.

The Popielno beavers were flying in. The Poles had agreed to provide two breeding pairs of adults and their three kits. I went to meet them on their arrival at Heathrow. I arrived early and waited in trepidation for the plane to land. While the experienced staff of the quarantine facility were utterly relaxed, I was desperate to check that the beavers were okay and to reassure their former owners that they were safe. Eventually a radio call informed us that the forklift carrying their crates was outside. A huge steel electric shutter was raised and the vehicle drove in. The crates were unloaded onto trolleys and, as the centre occasionally dealt with seals, transferred to two secure, tilled enclosures with pools in the main building.

Once they were unloaded with a pair in each room, I cautiously raised the slide on the first and bent forward to peer inside. Nothing was visible other than a large pile of straw, which I stepped forward to part with care. Underneath, curled tightly into balls, were two adult beavers. Their eyes were shut and it was difficult to determine whether they were breathing or not. I touched and stroked the nearest, whose silver aluminum ear tag informed was a female.

Nothing happened.

Convinced that they were dead, I crawled in further on my knees to get a better look while the quarantine staff peered cautiously from outside. As I reached out again to touch her, she let out a low hissing growl and, turning with incredible speed, jumped towards me with her teeth exposed. I leapt backwards and in doing so banged my head hard on the crate roof before falling out onto the tiled floor beyond.

The quarantine staff stampeded in a herd for the exit door, which they slammed promptly in their wake. They peered with pickled onion faces pressed tight to the door's misty glass plates as the beaver bounced towards me on the floor, hissing. As I scrabbled backwards on my bottom, I grabbed an abandoned yard brush, and as she grabbed my left boot and readied to bite, I stuffed the bristled head straight into her face. Like lightning she grasped it, leant over the bristles and with a single bite severed its head from the shaft. As I skittered further back trying to gain my feet, she pursued me vigorously, reducing the length of the remaining pole with a series of savage, lunging bites. I reached the wall, stood up and realised I was bleeding quite profusely from a cut on the top of my head that was making sight difficult and, just as the beaver launched another bounding, biting barrage, fell sideways into the seal pool.

I have no real memory from this point onwards of exactly what happened next. The beaver jumped into the water beside me and then, when the advantage was completely hers, decided that a relaxing bath was really much more appealing.

Slipping, sliding and blood-stained, I lurched towards the door wailing and fell through face forwards onto the warm wet tiles of the floor.

Although caring, the staff failed emphatically in their efforts not to laugh.

I did not find it funny.

I called her 'Grumpy' beaver. She was never amicable. Years later when she died and was post-mortemed, we realised that she had a grumbling set of stomach ulcers that went someway at least to explaining her general disaffection.

Before she died, she led a life of some adventure. In 1998, the sale of the New Forest Zoo forced the removal of the by-then fine collection of British mammals I had assembled that were unwanted by its new owners. A friend's estate in Sussex offered temporary sanctuary while I tried to find another site to restart the breeding centre for native mammals that I wished to create. The beavers came back from Blackpool Zoo when their quarantine had finished. Grumpy came with them. She was always good for a fight. It mattered not how quietly you tried to drain the galvanised swimming tank in her enclosure, she always heard you coming and would come grumbling out ready to attack. One day she raced out of her lodge with more than her usual vigour, and as I ran for the fence she took a series of bouncing bounds across her enclosure before leaping from some distance

for its bath headfirst. As I had just drained it for cleaning, rather than landing with a splash in its depths, she
smacked face first into its base with a resounding 'dong'.
Needless to say, this did not improve her humour.

The enclosures we had created to contain the beavers
were constructed from sheets of 4-foot-high tin, bolted
onto metal posts on their outside. Although a novel
design, their walls were gnaw-proof. While they seemed
tall enough as beavers are not jumpers, I did not realise
they could move materials. Grumpy beaver did. One
morning when we came in to feed her, she had pushed
a gigantic pile of wood bark and straw into a ramp-like
structure in a corner. She had climbed up this before
falling off its end and vanishing into Sussex. To say when
we discovered her loss that our subsequent search was
a needle-in-a-haystack exercise was an understatement.

There was a hell of a lot of good, wet woodland habitat
out there. We looked and found nothing. My girlfriend at
the time, who was never strong on logic or the repercus-
sions likely to stem from her lack of it, insisted that she
was going to inform the police. As she proved obdurately
impossible to dissuade, I decided not to join her in the
station and was therefore not present when she told the
responsible desk constable that she had lost her beaver.
I could however hear the howls of laughter that ensued
from the pavement outside.

In the end, Grumpy did not disappear. Months later a
local farmer who noticed that his cows were spending an
unusual amount of time in a single corner of their field

staring into a ditch went to investigate and found a large black creature building a pool with sticks. He called the RSPCA who sent out an officer. After eventually capturing the mystery beast, he identified it positively as a fat otter. It was then sent to Drusillas Park, which, having kept Canadian beavers for years, felt sure that it was a coypu. Eventually it was returned to us by an artist friend, Stella Quayle, who had trained her own small colony of Canadian beavers in the wildlife sanctuary she ran to march out in line daily at 3 p.m. to take carrots gently from her hands. Although Stella knew Grumpy was different, she also knew she was a beaver and returned her to me.

The beavers – which spent the night after their arrival swimming, feeding and reducing the sides of their transport crates to sawdust – were all, including Grumpy, in bed by the morning. None the worse they were transferred to their quarantine facility where they settled in well.

In 1999, when their six-month quarantine at Blackpool Zoo was over, they came to a new wildlife centre I had been employed to manage in Kent. The established family settled into an enclosure with a large central pool while Grumpy beaver, whose mate had died, was interred nearby in a smaller lagoon. I began to talk more widely about beaver reintroduction in England and a soft surge of media and public interest began to swell. One early visitor was John McAllister, the then East Kent–area warden for the Kent Wildlife Trust. While his bouffant crest of curly hair caused a whole heap of drunken Norwegian teaching staff – when we went there to catch beavers – to believe, to his great perplexity, our canard that, yes, he was Leo Sayer incognito on a winter getaway, his acerbic wit did not always endear. He was, however, determined that beavers could play a part in rewilding part of the Wildlife Trust's wider landholding. His zeal and grit were the loadstone of what followed.

The Trust owned the last fragment of old Kentish wetland – a frighteningly tiny morsel called Ham Fen, ironically located to the south of the historic cinque port of Sandwich – which they could only manage with difficulty due to its inaccessible nature. The site contained a range of rare plants and insects in its 9-acre core that was surrounded by a much larger acreage of 'improved' land that had been reclaimed over time for farming. *Improved* means quite simply that you smash, poison or reshape marginal land to serve a production goal for which it is generally ill suited. Then you insert the drains, apply herbicides or chemical fertilisers and receive vast grants for eviscerating habitats that possibly date back in time to the Bronze Age and replace them with lifeless monocultures of Italian ryegrass. While this process generally delivers little in terms of produce, intricate communities of fragile wildflowers, herbs, insects, fungi, birds and other small creatures are destroyed on a staggering scale to achieve its common end.

Although the fens centre was designated as a National Nature Reserve, no sustainable technique or process the Wildlife Trust might devise could address its issues. In its middle, tree growth was shading out the old fen flora. In the pasture, ryegrass dominated over a peat soil mineralised to artificial richness by fertilisers. It was drying out rapidly and, with the loss of ground and surface water bodies, any prospect of restoration was fading. Could beavers help? While their tree-felling abilities are well known, what is less commonly appreciated is that they always like to be close to water. In most countries, studies demonstrate that the bulk of their feeding activities are within a 10-meter radius of the water's edge. This figure undersells the extent of their impact as, once beavers have built one dam to create the first pool they require, they immediately start to excavate a complex system of others linked by shallow canals. These extend out into the surrounding environment from wherever the water's edge may be. They block them when wished with further impoundments and eventually what was once a tree canopy covering a watercourse opens complete to the sun as a complex wet, woody Venice affording a more than ample supply of forage around its greatly extended rim.

John's proposal was to put beavers into the fen to fell the trees and to raise the water levels, but both of us knew a licence application to simply release them would fail. The opposition from influential interests and individuals was just too great. Sure, we could fill in the silly forms. Our scientific argument would of course be sound. We might research a history and comply in triplicate with international guidelines. Experts overseas would support. There would be positive press. English

Nature might even feel brave enough to summon some soothing vowels.

All would be to no avail.

Some ignorantly obdurate old men would meet quietly with a brandy in a London club and agree with the linkage of a few crisp words that it could not and would not be done. That's how democracy works. We would be wasting our time.

Was there another way? Together we looked at the legislation. As beavers were not a dangerous wild animal, you needed no licence to keep them, and if we were not releasing them into the wild, no permit to do so was likewise applicable. If we put a fence around the reserve with water gates in its main stream, we could contain them in a pen of approximately 100 acres. Because Ham Fen was a National Nature Reserve, getting formal consent from English Nature to put beavers in was going to be a consideration, but Dr Grahame Bathe proved quite willing to help. This, at least, would not be a problem; he was keen to see progress. The Trust approached a sponsor for the fencing who said yes, and as the thinking at that time was that Norwegian beavers were nearer in type to the form that once existed in Britain, we arranged to obtain these from Dr Frank Rosell at the Telemark University College, who was studying them in detail at that time. Frank is an amazing man. Blonde and blue eyed, he has researched pretty much everything from cave bears' territorial claw-marking systems to the inside linings of dogs' snouts. Internationally respected, he is a firm, fit, typical Norwegian with one exception. He can't stand the sight of blood. Especially his own. This last characteristic might be overlooked in normal society, but in the high-octane hunting culture of Norway where gore abounds quite normally, it's a bit odd. Despite this – and his complete lack of a sense of humour – he is a good bloke.

Norwegian winters are severe, and some of his beaver-study families were living in suboptimal habitats, where Frank felt sure they were unlikely to survive. These were the ones he would catch. Frank's team used speedboats and spot lamps in the subfreezing temperatures of the Norwegian night to capture the beavers in the spring of 2001. In the crystalline rivers where the beavers lived, they could readily be spotted swimming underwater, and when they surfaced after a short chase in the shallows, they were netted by field workers who launched themselves like skilled seals into the icy water to do so. It was impressive to behold.

We built a new quarantine facility and applied for an import licence for twelve beavers to provide two family groups for Ham Fen and some other additionals for projects in Scotland. The fencing began and the BBC sent a film crew to Norway to record their capture. There was significant media interest in the story, the thread of which in general was entirely positive. All was going well.

Then a phone call came from the Ministry of Agriculture, Fisheries and Food (MAFF) – which, in an acronym extravaganza, was forced to change its name shortly after having finished burning a lot of Britain's cows and sheep badly during a gargantuan outbreak of foot-and-mouth disease to the Department for Environment, Food and Rural Affairs (DEFRA) – who claimed authority. They said they had full jurisdiction regarding the project and that further licences would be required. They could provide no clarity with regard to what these might be but made it quite clear that they would not be easy to obtain and that they were not, in principle, supportive. They would give no guidance with regard to what had to be done and no indication of an appropriate time scale. They asked if we would be willing to return the beavers to Norway. We said no. They hinted they might have them killed. We said no to that as well. When the Trust tried to directly contact the then Secretary of State for the Environment, Michael Meacher, to explain this predicament no response was forthcoming. The project had stalled.

As this bureaucratic chess game played out, I got a call one morning from an obscure government committee called ACRE whose normal remit was to advise on the introduction of biological controls for invasive insects in the horticultural industry. They had been tasked with giving an expert opinion on the beaver project. Would it be convenient if one of their representatives called to see me the following week? The duly appointed date arrived and a small man in a brown serge suit with a large brief-case was delivered by taxi. This creature wanted to see the beavers. After expressing surprise at their size, wincing a little when they growled and confirming that he absolutely did not want to step into their enclosure to get a closer look, we had nothing much else to discuss. After a few final perfunctory questions, he was gone. Three hours later a slightly more animated version of the same individual called to explain that he had left his briefcase in my office by accident, and would it be possible to get it sent back by motorcycle courier to London immediately? I complied with his request completely. After photocopying its contents.

Amongst other treasures the briefcase contained was the advice from DEFRA's legal representatives, which made it clear that they had no jurisdiction whatsoever in the matter of enclosed beavers. It was all a gigantic bluff but by this time the trustees of Kent Wildlife Trust were thoroughly rattled. The threats from officials of unlimited personal fines if beavers escaped, coupled with a mirage of potential jail sentences, had worked. They wanted no fight and were flying what white underwear they could assemble high. Compliance for them was the only option and the evidence of chicanery would have to wait for another day.

Further correspondence with senior civil servants ensued, but they were cold and non-committal. It suited their purpose to play for time.

Then the inevitable happened. The beavers, which had been quarantined for thirteen months, started to fade. We could not provide them with sufficient activity in their indoor holding pens to keep them fit. Alive for a time, yes. Fit, no.

One died. Then another.

An emergency meeting was held and an appeal sent to the civil servants. When no response was received, for me there was only a single option left. To photograph the dead beavers being post-mortemed, call the press and inform the Norwegian Ambassador that his country's gift was dissolving for no reason in front of his eyes. This was not a position agreed upon by all those attending, but with the consent of my boss, I spoke to the BBC One O'clock News, who agreed to run the story the following day.

It produced an almost instant response. Confronted with a potential diplomatic incident, our flustered opponents were quick to placate. Gone was the lofty arrogance of obstruction. It was all a misunderstanding. Could we just let the media know? Of course, the minister would see a delegation which would, prior to the meeting, have to agree to abide with whatever decision he reached without dissent. I was not allowed to attend but it could be organised quickly. John went to talk with the Trust's chairman. Meacher expressed surprise and claimed no knowledge of our problems.

Of course, it was possible.

What a good idea.

Sullen bureaucrats were blamed and permission was given to proceed.

We released two pairs and their single kits that had been born in captivity into the reserve with no fanfare shortly thereafter. They died after a short time, shrunken and flaccid, ruined by their long captivity. In time more from Poland and Bavaria were obtained and these established well. They coppiced the trees in the old fen centre, and the vegetation responded with a natural, vital vigour to their gardening that opened its heart to the sun. Their dams slowed the flow of the site's main drain and spread out the water once more onto the land. They created complexity and became its throbbing pulse. It was the first project of its kind.

CHAPTER FIVE

If You Look Now at Her Face

W HILE SINCE THE EARLY 2000S I AND OTHERS had provided some beavers for zoos and enclosures such as Ham Fen and that developed by the brilliant naturalist Sir John Lister-Kaye at his Aigas field centre on the Black Isle, we were doing no more than lighting very small candles of hope in the dark. Although there was a lot of talk, any prospect of a free-living beaver reintroduction back into the countryside seemed very far away. In 2007, a proposal to release beavers in Argyll was turned down by an embittered civil servant who had been passed over for promotion. It was not what was expected but it delighted those who had opposed from the beginning the mere idea of anyone even thinking about beaver restoration.

Most memorably, I met one of their number at the Scottish Game Fair in 2002 when I took two young female beavers for live display to explain to visitors something of their lifestyle and ecology. Now for those used to landscape politics, the expectation that the audience was going to be solidly hostile was misplaced. Although one of the staff in the Royal Society for the Protection of Birds (RSPB) tent nearby had been punched in the face the day before for simply pointing out that killing protected birds of prey was probably wrong, no one had come close to punching me and for the most part very few were greatly troubled by the prospect of wild beavers. A minority of them expressed a positive view and approximately the same proportion a negative one.

The difference with the latter was the basis for their viewpoint, which was crystallised for me quite clearly by a single individual. Massive and beige in a wool tweed suit, he was the father of one of the chaps with whom I was working. On the day we spoke, adorned with a fore and aft 'deerstalker' hat, he came striding towards me clasping a ram's horn crook in his meaty hand while his son simpered like a puppy behind him.

Humphrey was a nice guy and I did not want to start an argument with his dad.

'Beavers', he barked. 'Bloody big rats. Why bloody beavers?'

Taking my cue from his attire, I replied 'Well sir, duck ponds. Reintroduce this species and it will make lots of ponds, enabling you to shoot more ducks.'

He did not exactly smile although his lips moved to form a grin of sorts. 'I hate you nature conservationists', he said. 'You're nasty! I am the best nature conservationist

I know. When I walk out onto my estate, I love to see an odd sparrow hawk, but I don't want to see too many as the bastards will be eating my pheasants! And seals!' Now he was getting into his stride. 'I love seals! When I go down to my estuary, I love to see a few but too many and they will be eating my bloody salmon.' By now in full flow, he served his obnoxious finale, 'And when I go to Birmingham it's full of coons. Wogs and coons!'

He did not speak of control, but I had had enough of his unpleasant rant and turned away. He left with a 'hippotic' harrumph at a determined pace, secure in the knowledge he was right. Humphrey shuffled off behind him, fawning weakly in the wake of his self-righteous cathedral of a parent. For fifteen years the objectionable bugger was and, despite progress on some fronts the Scottish Government, is still dancing a jig to the tune he and others of his Pleistocene choose to play. Although a tiny minority their influence and ability to manipulate a political outcome despite science, intelligence or right is profoundly shocking.

———

Martin Gaywood is tall, stick thin and gawky. Although a professional public servant who works for SNH, those of us who have known him for years don't hold that against him. Like many biologists he has subsumed the characteristics of his earliest study species, which in his case is unfortunate as that species is the smooth snake. Smooth snakes are both rare and dull. They exist in heathland environments in southern England, are rare because of habitat and prey loss and are generally patterned slate-blue on their upper body with orangish bellies. While Martin's belly is as far as we are aware not orangish, his tongue under stress does have an unfortunate tendency to dart in and out in a reptilian fashion in uncertain social situations.

Martin was and still is largely responsible for beaver reintroduction in Scotland. In an age when many who work in professional nature conservation only do so to make their mums happy, Martin for his creed has shown a remarkable spirit of perseverance. We appreciate that he has had to fight many battles on his own.

In 1999, I was with him in the ancient forest of Bialowieza on the Polish/Belarusian border at a conference on beaver reintroduction. Although well known for its European bison or wisent, Bialowieza also has free-living beavers that were reintroduced to the forest. The conference was the second European Beaver Symposium – an event that occurs every third year in a different host nation – and was on this occasion held in Poland. Scientists and field workers from all over Europe and a few from North America gathered to give presentations about different aspects of beaver history, biology, ecology and management.

Get-togethers of this sort are generally great fun and this was no exception. Huge Finns with Santa Claus beards rubbed shoulders with dapper Swiss attendees and tidy Danes from their state forest service who have to wear comedy costumes when their king comes to hunt. Poles with gigantic moustaches drank tinctures of many colours and smiled. The French talked amongst themselves and were rude while most of the eastern Europeans gave tediously dusty presentations before getting very drunk on the vodka they concealed in the inner linings of their coat pockets. An elegant lady from Argentina explained how the introduced Canadian beavers in Tierra del Fuego were stripping valleys bare of southern beech, which, unlike the riparian tree species that through long association have adapted to accommodate beaver feeding in their native landscapes, is not.

A small scientist from Oxford University gave a splendidly pointless presentation on why a programme of beaver habitat modelling, which her faculty had developed, would ensure that in Britain, when we eventually got round to reintroducing a species everyone else had got round to reintroducing fifty years before, it would be the best reintroduction ever. It says much for the courtesy of her international audience of vast experience that even a few of them bothered to clap.

One merry night after the Russian contingent had drunk massive amounts of beer and vodka and sung 'Katyusha' with glass-banging blows to their table, a small band of English foresters responded with a reedy rendition of 'On Ilkla Moor Baht at'. They tried to encourage those of us from other parts of the UK to join in, but as it was quite, quite clear by the time they had sung the first word or two that this course of action was a massive cultural mistake we felt it best with a degree of glee to observe rather than participate. Some of the Welsh chaps went to the toilet. We could hear them laughing uncontrollably inside. When the song finished no one cheered and most observers went to bed. The singers shuffled off the stage into the shadows and thereafter maintained a low-level social presence. It was a queasy experience, which twenty years later many of those attending still remember. It's likely that some will carry the mental scar to their grave.

Most of the many beaver ecologists and enthusiasts working with the species in those early days were colourful. None more so than Olivier Rubbers. Olivier from Belgium was tall, elegant and aggressive with cropped dark hair. At the time of the Bialowieza conference, he headed up a Belgian organisation that was similar in sort to a militant Greenpeace. It fought polluting companies, farmers' organisations, inadequate government and its own membership alike. Wherever there was a fight in Belgium regarding an environmental issue, Olivier was there and he was angry.

Unsurprisingly, the issue of reintroducing beavers – which had been extinct in Belgium since 1848 – caused conflict. It was and is Olivier's contention that he tried rationally to discuss the return of this species with both the government and its farming allies who said no. It is their contention that he did not. In any case what then transpired was splendid.

They said no and he said 'Fuck you!'

Together with a band of co-conspirators, he drove to Bavaria where beavers were at that time freely available, acquired one hundred and drove them back to sites in the Ardennes they had pre-selected as near ideal. At 2 o'clock in the morning they phoned each other, let them go simultaneously and then went home to bed. The morning after, an astounded nation woke to the news that beavers had been reintroduced.

Olivier's actions and the positive response they received inspired people to kiss him in the street. While the general gist was favourable, the policemen who arrived at his home later in the day to take statements and to confiscate his passport were not impressed.

The government was furious. They consulted their statute books and discovered that no laws existed in Belgium regarding who or how an extinct species could be reintroduced. Despite this they maintained it was an illegal act. A wandering beaver from the released population turned up in a back garden in Brussels and was captured by a government coypu trapper. Belgian zoos declined to afford it a home. The government phoned the Bavarians who told them they accepted 'no returns of secondhand or otherwise used beavers.' In the end the Belgian Government released it, confounding their own position but starting a long running legal pursuit of Olivier that lasted for around five years.

Meanwhile, the released beavers bred successfully and expanded their range. Scientists studying the impact of the wetlands they were creating in great profusion in the wooded valleys of the Ardennes observed that their dam systems slowed the winter surges of water so significantly that they prevented downstream communities and infrastructure from flooding. Other beavers moved into Belgium from a population in the forest of the German Eifel. Tourist excursions running 'Beavers and Belgian Beer' tours from Brussels used local hostelries to observe both the beavers and the burgeoning population of rare species, such as black storks, which were returning to reoccupy the habitats they created. There are now more than eighty breeding pairs of this once-extinct bird in Belgium.

At Olivier's final court hearing, when the government pressed for all liability to be placed on him personally for any past, current or future beaver damages, his consul pointed out to the judge that even if he were to accept responsibility (which he had no intention of doing) then it could only reasonably be for the Bavarian beavers that he 'might' have released and their descendants – not for the interlopers from the Eifel. As he explained that proof would therefore have to be provided with regard to the origins of the beaver responsible for the felled tree in question, the judge smiled, stopped him and declared the entire exercise a waste of court time. Olivier was fined a total of 500 euros for the transport of a protected species without a licence, thus making his reintroduction one of the cheapest of all time.

After he told his story in the long, pine bar in Bialowieza, he turned to Martin Gaywood and said, 'You people in the UK have talked for too long. Here's what I will do for you. Give me your address. I will come to see you. You will not know when. It will be a secret but one night when you are lying in your bed you will hear a knock on your door. You will get up, walk to the window and draw back your curtains. Outside will be a van with Belgium number plates and inside that, well . . . I will have a little surprise for you!'

Martin tittered and gulped so hard he almost swallowed his Adam's apple. His tongue began to dart and shortly afterwards he disappeared off to bed.

———

It was in Bialowieza that I first met Gerhard Schwab. For those who have never met him, Gerhard is a Teutonically huge figure. Big voice, big hair, big beard, big tummy, big heart but small, small dog. He looks grim and gruff but for those of us who know him well and have seen him play with baby beavers or tell respectful hunters that the beaver is 'mien liebling', he has become a living legend. His story of his own personal journey with the beaver has become a benchmark in a time when the issue of how we redevelop and define our future relationship with wild creatures has never been more acute.

Gerhard became involved with beavers in Bavaria at a time of considerable tension. In 1966, when the species was first introduced, beavers were hard to come by. In our age of indifference, we forget that in the exhaustion that followed the aftermath of World War II the concept of nature restoration was hugely appealing.

It was healing.

It was rebirth.

That beavers lingered long on the Danube is well illustrated by a remarkable document from the early 1800s I was shown by a Bavarian wildcat expert. In meticulous detail, in tiny, hand-coloured pen-and-ink illustrations, was a recent world that has been painfully, achingly lost. Through its wide grasslands the now near-extinct field hamster was scattered in abundance. Eagles soared, wildcats lurked, vultures of several species sat aloft on their miniature Alps. Lynx and boar roamed forests in the darkest depths of which lay the lairs of the grey wolf and the great brown bear. In the river lands, scattered beavers felled and fed in freedom before the last surge of slaughter claimed them all. By 1867 they were gone.

Although exact figures are unclear, by the late 1900s the beavers that remained had been reduced to approximately 1200 individuals scattered through France, Norway, Germany, Poland and Russia.[1] Smaller populations persisted elsewhere in Mongolia, China, Turkey and possibly Iran.

In odd parody it was the hunter's association that brought them back in the early 1960s. Although strict protections from the beginning of the twentieth century onwards had enabled a rise in numbers of the surviving populations, and reintroductions from these had been undertaken when their numbers allowed, they were still at that time not considered to be an abundant species.

When the East Germans refused to play ball with their decadent western cousins, beavers were obtained from the more cooperative Russians instead. The first brought to Bavaria were released initially into enclosures that were built to surround old gravel pits and oxbows on the Danube. After the first escapes – two days after the first releases – other beaver releases followed onto the main river itself. While initial concern focused on their ability to survive on such a large engineered watercourse with concrete- or stone-lined banks, their presence was eventually forgotten when they vanished and anyone who remembered considered them gone. A failed experiment. A dream perhaps that was just too far.

But decades later odd things started to happen. Large stick nests – which were obviously engineered out of timber, mud, roots and rocks – appeared stuck to the walls of the Danube. Dams built out of maize stalks appeared in drainage ditches. Farm equipment collapsed into voids in fields that had been cultivated without issue for a generation. Tractors sank in water-logged soils. Tidy patches of sugar beet were removed from croplands by robbers, who followed well-worn paths to and from the river's edge. Ornamental trees in parks and gardens fell down. Encounter and observation over time confirmed that beavers were the culprits.

Ordinary people and the nature conservation groups were enchanted. The beaver, the bringer of life and nature was returning to a landscape near them. Fantastic! The farmers on the other hand were less enthusiastic. It was their sugar beet that had gone. Their tractor in a swamp. Their cultivator with a newly U-shaped axle.

The farmers spoke to their leaders who agreed. Beavers were bad and had to go. They were standing in the way of efficient food production. The nation might starve. When others laughed, they got angry, and as protagonists insisted stridently that the protected beavers must not be harmed, their illegal killing began. Lodges were burnt out. Beavers were shot and snared.

In Belgium, where the same tensions have arisen, individual beavers have been captured in live traps and crucified. Alive. Some individuals' depth of hatred for creatures whose abilities defy or transcend theirs is primal.

Into this climate of distrust, hostility and extreme polarisation strode Gerhard. As a recent master's graduate in wildlife management from Colorado State University, it was his job to sort it out. He travelled widely to see issues first hand, spoke to conservationists, farmers groups, communities and individuals alike. He designed solutions that defused tensions for a time, such as electric fencing systems for cropland protection, which gave the bumbling thieves with big button noses the literal 'shock of their lives' when they sniffed its hotwired strands. Dam-drainage systems were devised that beavers could not block but that lowered their water levels easily. He constructed traps, wrote management guidelines and persuaded recalcitrant funders to support the trapping and translocation of beavers in problem areas of the croplands to other countries where they were extinct.

This well-organised process provided foundation populations for Croatia, Romania, Hungary, Belgium, Serbia, Bosnia-Herzegovina, Spain and Mongolia. By coordinating this exodus, Gerhard saved the lives of nearly a thousand beavers. The beavers responded by expanding to fill the watercourses of their new environments with progeny full to the brim. There are many in all these nations now. But in the end once this option had achieved all it could and there was simply no space left, nowhere to move them to, nowhere to go, then the killing had to come.

The problem is simple. When beavers were abundant after the last ice age, the landscape was theirs and we were either absent or present only in very low numbers. Humans could not engineer water in the way beavers so capably could. We clung to their coattails as hunters. When we started to farm, we took and then flensed the water lands. We built our civilisations in landscapes that once belonged to them.

Some cultures understood their function well. In ancient Iran, beavers were one of the most sacred of creatures. As they were believed to be formed from the ghosts of dogs, they were called 'water-dogs'. A single beaver was thought to have as much holiness as a thousand dogs, and it was believed that killing a beaver would produce drought, which would cease the growth of the crops until its killer received punishment. Those who harmed beavers had to pay a heavy fine of 60,000 dirhams and kill 10,000 snakes or tortoises to compensate for their sin.[2]

Although modern science demonstrates the truth of this understanding – with the requirement to kill the snakes and tortoises being the only debatable part – in the arid lands of North America, we, at a time when our dominance of the planet is near complete, require them to dance to the tune we set. Rough wetlands at the edges of farms; some state forest blocks that were always damp; land owned by nature conservation bodies; in the woody fringes of watercourses when we don't want the trees anymore. Where there is space, then no conflict may arise, but when they flood areas that are flat and fertile, when a road goes underwater or burrow systems are punched into flood walls, we say no and they must die.

Gerhard understood this, and despite considerable opposition from disgruntled nature conservationists, he pioneered through experience of all sorts an approach to beaver management that is pragmatic and humane. His system is now internationally respected and forms the basis of beaver-management programmes in other cultured countries. One day when we in the UK grow up and begin to restore beavers at scale, it will form the basis for ours.

Bavaria, if not a second-home, has at least become a place of which I have become very fond. I have encountered so much kindness there. In Gerhard's time, with good will and collaboration in some communities at least, the beaver has transitioned from being simply a problem to a position of much greater respect. Some villages such as Pforring near Munich hold an annual community party where beaver-shaped pretzels and beaver-tailed bread vie with beaver beer or Bibergiel schnapps (it's horrid) for positions of novelty.

The old wooden barn decorated with corn dollies, the skins of wild boar, stuffed stone martens, old farming implements and other sundry curious bygones has hosted many events of this sort. When they begin on warm summer evenings, they do so to the wheeze of accordions and the synchronised symmetry of sweet-singing men's choirs. Although the music becomes more contemporary when evening meets morning and a combination of sausage and schnapps takes its toll, the traditional Bavarian hangover cure of broiled Wiesswurst and Wiesbier is guaranteed to make you feel, if not well, at least very different when you awake.

During the years that I have known Gerhard, we have organised many tours of interested groups from Britain to view his exceptional work. One of the first was for a mixed group of naturalists. Some were frisky and fun. Others less so. Gerhard's sense of objectivity, which at times can be unthinkingly pragmatic, reached a peak of dislocation when he demonstrated to a rather prim Scottish lady how to sex a beaver. Now – beavers have no visible sex organs, so unless a female is suckling and you can see her exposed teats it is impossible to tell gender from field observation. To do so you must first catch your beaver. If it's a small or middle-size one, they will have folds of skin down each side. Beavers are not generally agile, and if you pin them down and grip them firmly tight behind their front legs, it might end well. If they are too big for this, it's best to use large, flat boards to restrict them into a corner of a holding pen and then to try to get them into a sack.

Sometimes you can't tell until you have caught them if something that looks like a candidate for the former category – small or middle size versus big – has, when you have pinned it down, so filled out with muscle that:

- It has no skin folds at all.
- There is nothing to hold on to.
- It's about to turn around within milliseconds and bite you hard as hell.

In the 1600s the author Edward Topsell observed that 'the bitinge of this beast is very deepe, being able to crash asunder the hardest bones, and commonly he neuer loseth his holde vntill he feeleth his teeth gnash one against another. Pliny and Solinus affirme, that the person so bitten cannot be cured, except he hear the crashing of the teeth'.[3]

It's therefore a relief that in the real world beavers very seldom bite people. In May 2013, *The Guardian* reported that a 60-year-old fisherman had died in Belarus after being bitten by a beaver that severed his femoral artery. Although the circumstances of this exceptional event are unclear, it is believed that he may have been drinking more than a little vodka before attempting to pick it up and kiss it.

The misconception that beavers keep their teeth sharp by gnawing is nonsense. By grinding their pairs of upper and lower incisors together in quiet moments, beavers sharpen their teeth on purpose. The hard outer, orange enamel, which is strengthened with iron, is backed by a softer white dentine that is eroded by this process to leave a razor edge so enduring that in the Bronze Age they were used by people as chisels.

I have only ever once been bitten by a beaver. It was in my left chest. It was horrid. As the beaver slipped from my grasp at a demonstration it turned, pressed its head into me and closed its opened mouth. I felt its teeth click when they met and then thankfully it let go. A note in our work's accident book recorded that 'Derek was bitten through left breast by beaver'. As a remedy it recommends that other staff 'Don't pick up 3-year-old beavers [as] they are too bloody big and strong'.

The beaver bit with inexorable crushing power through my jacket, top and T-shirt. Surprisingly, the wound it created did not bleed much but it did go septic when the ends closed over, as most beaver bites do. Opening the ends and syringing out the contents daily with a sterile fluid was about as much fun as it sounds. I do not want to be bitten again.

However, providing all the forgoing doesn't happen and you have subdued your beaver successfully, then an assistant is required to sex it for you. That individual will ideally have small, slim fingers. But if not, any size of functional digits will do although big ones with roughened skin are not good. Aiming for the beaver's pink, quivering cloacae, you open its lips and by extending your fingers inside try to feel for a small oval gland located in the inner wall. It's pretty slimy, but once found it should be gripped and then protruded from its fleshy retreat. At its top it will have a small nipple tipped with a black dot and occasionally a couple of hairs. With two fingers of one hand holding it as firmly as circumstances allow, you then need to squeeze it firmly with two fingers of the other hand.

You must be careful amongst the fluids that will be flowing by this point to focus only on that coming out of the nipple itself. If successful a female will produce a toothpaste-type grey exude while male excreta will be viscous yellow.

Although beavers are biologically incapable of facial expression, they don't enjoy this process much, and while it's essential for us to know their sex for a variety of reasons, they are almost never grateful. When you put them down after the procedure is over, it's best to step away fast, or better still to have your assistant place a stout board between them and your lower legs.

While most stumble away without reluctance others will bounce swiftly back on their hindlegs, with their incisors exposed ready, if possible, to exact a bloody revenge.

A procedure of the sort described above was what in essence we tried to achieve for our first training tour. We had caught an unwilling candidate that I was holding while Gerhard attempted to obtain a result. The Scottish lady was standing in the middle of a throng of onlookers; she stepped closer to get a better look.

Gerhard inserted his fingers and withdrew them swiftly with a chipolata-size lump of nut-brown, woody material. That he said 'is shit'. He flicked it downwards as the audience watched, and tried again. His next attempt was rewarded with a stream of urine, most of which splashed straight onto the Scottish lady's neatly polished shoes. Observing 'that's piss,' Gerhard continued.

A chalky yellow runnel of castoreum then followed, before finally grasping the gland firmly in his mighty paw Gerhard extracted a tiny, toothpaste-like squiggle of grey. He turned with it on his thumbnail at exactly the moment the woman leant forward and by doing so inadvertently stuck it to the tip of her nose.

'See,' he said, pointing to the rest of the group, 'if you look now at her face, you can see it's a female!'

CHAPTER SIX

A Disabled Walrus Might Be Possible

I N 2003, I MOVED TO DEVON TO FARM AND AS AN ASIDE created a small quarantine facility for beavers in one of my outbuildings. When required I could, if I wished, import beavers for zoos. I travelled to countries such as Romania where shoelessly poor forest officials were reintroducing beavers. I witnessed beavers in Cornish-type landscapes in Brittany. I went to Holland to see them reintroduced into Dutch dams and dykes, entirely artificial, completely man-made environments. I began to understand how beavers lived and what they could do.

In 2004, Dr Simon Pickering, the then CEO of the Cotswold Water Park near Cirencester, contacted me. He wanted to discuss beavers. Simon has a bald head, a gigantic brown beard and twinkling eyes. He is the sort of naturalist who would not have been out of place on a Shackleton expedition killing a narwhal with an axe in order to measure the length of its penis. He is funny.

When I arrived at his office, he was there to greet me in the company of a tall, languid, chisel-chinned individual called Will Vickery who had recently returned from farming in Romania. The subject they wished to discuss was a project Will was managing on an estate next to the water park called Lower Mill. Basically, Will's boss, Jeremy Paxton, had obtained permission to develop a luxury holiday home complex around a series of old gravel pits on this land on the basis that 75 percent of the remainder, once the principal development works were complete, would be managed for nature. He had asked Will for ideas, and in a bit of a blank moment Will had suggested they introduce beavers, which he knew next to nothing about. When Jeremy asked for more information, Will made it up, but the prospect of beavers building a home for nature so chimed as an advert for Jeremy's building company that he was told to proceed without delay. Could I help?

We went to see the site. The area proposed for enclosure turned out to be a series of steep-banked, well-wooded gravel pits. Will asked what had happened at Ham Fen and I explained. At the end I added that we knew that DEFRA's position was a bluff and that their legal advice was that they had no jurisdiction. He asked how I knew and I passed him the opinion of their lawyers. I heard nothing back for a week or so before Jeremy called. I explained the story again. 'Okay,' he said, 'organise an import of two families plus some spares. I will cover all costs.' I asked what I should do regarding DEFRA, and he said that he alone would address that issue.

A Disabled Walrus Might Be Possible

Following established precedent, we applied for import and quarantine licences and obtained the beavers from Gerhard. We quarantined them for their required 6-month period and obtained the final document to conclude the statutory process at the end of this period, which was a release licence from quarantine. Now for anyone getting confused by the utter tedium of this paper trail, this simply allows any incoming imported animals to proceed to their next destination when their rabies-monitoring period is over. It does not allow permission to release 'Not Normally Native' species into the countryside, which is what beavers are still legally considered to be, although no one really knows what exactly this term means. It has never been tested in law and has been broadly regarded as being the great big cork in the bottle of ambition that stoppers those who seek to restore Britain's impoverished wildlife.

Now DEFRA is an arm of government with a very broad remit. While its resource-strapped empire is responsible for ensuring that pig movement licences are issued in an orderly fashion in Pontefract, it is also concerned that any level of acoustic noise emitted from boom-boxes at kids' birthday parties in Broadstairs does not breach international conventions. Although a few fine people do exist within its bounds, most of its employees are drawn from the university of prevarication. Once employed their principal goal is to outcompete their fellows in ensuring pointless delay. The sort of fun folk it's a joy to spend Christmas with when the time comes to take the Sellotape off the shortbread. Although the situation has now changed, at that time DEFRA's various arms did not talk to each other. So those responsible for importing beavers who were only interested in their compliance with rabies-quarantine legislation were completely unconnected to those responsible for frustrating their release into the English landscape.

A date was arranged for the required two pairs to go to Lower Mill. I drove them there myself. Jeremy had said that he was going to do a press call but at no stage did I realise just how significant that would be. When I arrived before dawn, there to greet the beavers was pretty much the assembled media of the world. The BBC, ITN, Sky, Channel 4 News and other more specialised broadcast interests were all in place with their aerials up and headlights blazing from mobile units out into the dark. There was radio. The papers were coming. It was a significant event because Jeremy's press release simply said that today was the day when beavers were being reintroduced to England.

Along with Simon and Roy Dennis, I was briefed to assist with press statements while my staff took care that the beavers were fine. The opening event was the release of one calm female onto an open lawn in order to afford excellent film and photo opportunities. Beavers can't run very fast on land and if no water is immediately adjacent you can simply guide them quietly with boards in any direction you wish. When all was ready, we moved her crate to its appointed location and raised the slide at one end. Completely unfazed and with little concern, she ambled out to sniff and feed on the fresh willow branches we had provided.

As I turned to begin my first interview, Jeremy caught me by the arm and said, 'Let Simon do it.' Before I could ask why he said, 'Follow me', and strode towards a wall in the mid distance. Behind it, sitting on the bonnet of his car, was a policeman. 'This is the chap you want to speak to', he said airily as he turned and left me to it.

'Mr Gow', said the policeman.

'Yes', I replied.

'DEFRA says you have no licence to release these beavers here today, and I want you to go now and put that one back into its box and then drive home.'

I explained that I had the licences to import, quarantine and release them from quarantine and that, as they were going simply into a pen, I needed no other licence. After asking if the beaver on the lawn – which by then had decided to have a short nap unperturbed by the lights of its audience – was going to bolt for freedom, and agreeing it was not, he asked for my licences, read them and then called his chief constable. I did not overhear the entire conversation, which involved confirmation that 'yes there were several licenses' and that 'yes they were issued by, stamped and signed by officials from DEFRA', I did hear a noticeable rise in the volume of the chief constable's response as his sentences got shorter. After a few more yeses the officer walked back towards me shaking his head, handed over the licences and said, 'The chief constable has just said you are to carry on. Those bastards in Bristol are always doing this sort of thing to us. Have a nice day playing with your beavers, Mr Gow. Goodbye.'

I went to find Jeremy to explain this outcome. By that time DEFRA, who had learnt from the early morning news what he was up to, was counter briefing against him. They told *The Sun* newspaper that his beavers were 'dodgy' and that if he released them, he would be potentially liable for unlimited fines, a five-year jail sentence and that the beavers would be shot. Jeremy, who by now was thoroughly enjoying himself in turn, told the media that he was naming the beavers Tony, Cheri, Gordon, Sarah, John and Pauline, after the then Labour government's top brass, and that Tony beaver would be the first to explode in a spray of brains when he was shot in the face.

The papers liked this very much.

By the late afternoon, when DEFRA accused him directly of contravening the Wildlife and Countryside Act, Jeremy's lawyers were ready and asked them to explain how they could claim jurisdiction regarding beavers in enclosures when their own internal legal advice was that they had none.

Rumour has it that those responsible were hauled in to see the minister and told they were to have nothing more to do with beavers ever again. They had to publish three public apologies over the course of the following two days stating how eager they were to work with Mr Paxton, what a wonderful project it was and how they were emphatically not anti-beaver. Before returning to their lair. Where they never linger long.

———

In the litany of dim communications I have received from officialdom regarding beavers in more than two decades, perhaps one of the finest is their most recent.

At the time of this writing, DEFRA has banned the import of any more beavers from Gerhard. While the reason given for this fatwa is that their import will bring to the UK an admittedly rather nasty tapeworm called *Echinococcus multilocularis* (EM), which is currently not believed to occur in Britain, the background to this move is slightly more opaque.

EM is transmissible to humans and once ingested can form clusters of cysts in the liver, heart and brain. Once infected it is untreatable without surgery, which is hazardous and can be ineffective. The normal life cycle of the parasite is such that it develops in the internal organs of rodents – commonly voles, although a very few beavers have been recorded with the condition – which cannot transmit it to any other species through physical contact themselves. It is only when an infected rodent dies or is killed and consumed by a canid (a fox, dog, jackal or wolf) that the tapeworm becomes active when its eggs are released out into the wider environment in its predator's faeces. The eggs are tiny and windblown. When they settle on vegetation and are consumed by a rodent again, the entire cycle repeats itself.

People can of course catch it very effectively from dogs, which is why the pet passport system that allows you to take your favourite furry pal on holiday with you requires their worming by an approved European vet before you return. An estimated 150,000 dogs – those that can be legally traced, and it's very likely that there are others – annually participate in this programme. Expert vets experienced with the system are clear that a percentage of the treatments supposed to have been undertaken of returning dogs are ineffective and therefore it is quite credible that a number of dogs could be traveling back to Britain with the disease every year. To sit in your house next to you watching TV while licking your face or farting wetly into your carpet in front of your 6-month-old baby. Under either scenario an infected domestic dog will offer you and others the opportunity to come into contact with their faeces on a daily basis.

If dogs therefore have not already done so, they will in time bring EM to Britain. This likelihood is recognised as high risk by . . . DEFRA.

In contrast, in a busy year Gerhard was importing perhaps twelve beavers annually. While it is not known what percentage may carry the disease, it is likely to be low and the blood test and or laparoscopic screening required during their quarantine period can determine whether they have the condition or not. If they do they are killed and burnt.

Bear in mind however that the beavers carrying the condition cannot transmit it to anything else until they die and are consumed. If they decompose and then degrade the disease dies with them. It's simply – even though the senior government vets in Wales after labyrinthine, uninformed debate plainly can't seem to grasp this essential fact – not transmissible in any other way. So even if you collect, dry, sugar and consume as much beaver poop as you like on a daily basis for breakfast for a year, you can't catch it. If you kiss your improperly wormed dog Dillon just once on your return from Bordeaux, you can.

When on the 24th of July, 2018, Michael Gove, the then Secretary of State for the Environment, released two beavers that had been imported from Bavaria and had been screened clear for this parasite through blood testing into a fenced enclosure in the Forest of Dean, it was to the acclaim of many and the extreme displeasure of a few. In those days of a tiny conservative government majority, the vote of every MP was critical for an administration trying to navigate the stormy shoals of Brexit. One of Gove's junior ministers, who disliked the idea of beavers and any landscape change that might favour nature in any way, tried to bully the FC, whose idea it was, into not doing so. They were told to manufacture a reason for aborting the whole idea. They refused.

Gove got wind of what was afoot and stopped her, but in the end, she persuaded DEFRA – whose own internal assessments had shown that any disease risks from beavers were minimal – to support a case for retrapping them on the basis that they were diseased with EM. It was her 'golden bullet' to muddy the water, confuse and delay. Although it was quite simply nonsense, the beavers were caught and replaced by others from the River Tay. DEFRA then banned the import of any further beavers – but not voles or exotic pet canids such as Arctic or silver foxes, which can also carry the disease – from countries where they considered EM prevalent. However, as the government's 25 Year Environment Plan welcomed their official return, DEFRA does not want to be seen as anti-beaver and has therefore proposed a compromise to enable imports from elsewhere. This generous offer currently allows them to come from:

Malta. An arid island in the Mediterranean. Its drinking water supply is drawn from ground water and desalination. Although 10,000 years ago it was a forested tropical paradise harbouring populations of pygmy elephants and hippos – which need water – together with gigantic dormice, it is unclear as to whether beavers were there then. What is clear is that there are none now, although there is an active colony of Beaver Scouts.

Ireland. The only part of the UK where there have never been any beavers naturally to the best of anyone's knowledge. Although respected wetland archaeologists still believe that credible evidence of their former presence might one day appear, none has been found to date. There are no beavers in zoos there, but Beaver Scouts do exist.

Svalbard. This is Norway's part of the Arctic. If you want to obtain polar bears, reindeer or Arctic foxes, then Svalbard might possibly be a great option. There are no beavers, although when the Norwegian biologist I confirmed this with was able to stop laughing and draw breath sufficient to speak, he did say

that perhaps a disabled walrus might be possible.
Svalbard has no known colonies of Beaver Scouts.

Finland. The Natural Resources Institute Finland reckons that the beaver population there might total somewhere in the region of 25,000 individuals. The problem is that only around 5,000 are Eurasian beavers with the remainder being Canadians brought in the early years of the twentieth century when nobody knew the species were distinct. While it is possible to tell them apart through genetic tests, this option brings with it a level of near imbecilic complication that no protagonist of beaver reintroduction has ever advanced. It's so breathtakingly moronic that I quite forgot to check on the availability of its Beaver Scouts.

Norway. Well yes there are Eurasian beavers in Norway. Many of them. It's not entirely clear that even these are all EM free in any case, but well done, DEFRA, for your endearing sincerity of purpose.

We have learnt a lot about importing beavers over time. Experience demonstrates that they can be kept quite simply and are pretty robust. Providing when you move them that their crates are well ventilated, are kept cool, have dry straw and are provisioned with many sweet-eating apples, they are generally fine. Statutory rabies quarantine consists of keeping them in a series of indoor pens in buildings that are steel lined, with locking doors leading out into a central service alley. Inside, small bales of straw are used to create open-ended diamond shapes that are then roofed with wooden pallets and covered with willow branches. Through a combination of feeding and manipulation, the beavers will take the willow and combine it with the wood shavings they receive daily to create perfect mimics of the lodges they would naturally build in the wild. Over time these structures in their pens become large and elaborate. They sever any branches that protrude into the soft inner chambers and transfer finer sticks into these, which they then split lengthwise in quantity to create deep, springy, warm beds.

These pens contain swimming tanks to enable them to exercise and maintain their soft lustrous fur in excellent condition. Beavers, like other aquatic rodents such as coypu, muskrats and water voles, have a coarse external guard hair and a much finer layer of underfur. In the damp, cold conditions these creatures occupy, their maintenance of their snug coats in a waterproof state is critical, and as a result they spend a significant amount of time grooming both themselves and other individuals in their family groups. A beaver that looks generally bedraggled is probably unwell for reasons that have little to do with its coat, while a healthy beaver will simply emerge from the water, shake itself vigorously and within a very short period of time appear entirely dry. Beavers defecate, urinate and in season mate in the water, although quarantine tanks are not large enough for this last function.

Beaver faeces in the wild are generally woody and dry. Although they smell frankly beaverish – a not unpleasant blend of turpentine and coal-tar – in captivity their substitute diet of fruit and vegetables (especially carrots) ensures that, after a period of a few days, the chunky orange soup that results is generally unappetising and best disposed of swiftly.

On a recent swimming excursion with my kids in a lower farm pool, which the beavers from our wetlands frequent, we were intrigued to find numbers of marshmallow-size balls of what looked like pond weed bobbing obviously on the surface. Intrigued as to their origin as we had never observed them before, I caught several, which when opened were unmistakably comprised – in their soft, scented centres – of shit. Woody shards. Musky smell. Definitely beaver.

While I was fascinated to observe how quickly their excrement, with its tough blend of lignins, was being returned by what transpired to be an algal mantle to digestible matter for insects and fish, I felt it prudent from that point on to insist that at least when swimming the kids kept their mouths shut.

Once imported beavers are comfortably ensconced in their quarters, very few problems generally arise. Occasionally bacterial infections can lead to sores on their tails. Some odd individuals contract fatal respiratory disorders. Fights are very uncommon in settled family groups, and deep bite wounds are only generally encountered when single individuals introduced to create breeding pairs simply decide that they don't love their prospective new mates. While sometimes they can be persuaded, on other occasions their dislikes prevail – one large female introduced to a slightly humbler male simply started to wall him up daily in a chamber in her lodge to the stage where he could not leave to feed or drink – and other more pliable partners have to be sourced.

Serious conditions are rare. One family developed a MRSA-type superbug. When the electronic tail tags to enable their location after release were fitted – a procedure we no longer undertake as they seldom last long and can result in significant suffering and injury – they died quickly from the infection that invaded their opened flesh. While their single large, sleek kit survived, its parents rotted to a point where euthanasia was the only humane option, despite intensive care and antibiotics.

One larger than normal import of sixteen individuals developed a stress-related condition that weakened their gut linings. They were beautiful beavers in the prime of their lives. Although it seemed unusual, we were not alarmed when the first one died. Within days a morning check of their pens revealed another two dead outside their lodges. I checked all the others in their beds, sleepy and warm, bagged the two corpses in plastic yellow hazard sacks for their trip to the pathology laboratory and got into my truck. Before I started the engine, their keeper called to say another three were dying, stretched out on the floor, choking, retching, gasping. Dead within minutes. We bagged those and took them, too, but the following day another passed and another thereafter. Antibiotics saved the others.

By the late 2000s we were still confined to keeping beavers in fenced enclosures. Again, we wanted to explore the possibility of more advanced ambition. The idea was to move them beyond fences, out of enclosures and into free-living environments.

As Britain's river systems commonly flow from high land to the sea without interconnecting with other watercourses on either side, and beavers are poor colonisers across land, then in theory at least it should be possible to find sites where beavers could be placed without committing to any wider process of national reintroduction. This would enable appropriate scientific assessment and a cultural re-understanding to develop to the point at which beavers would then become a normal part of the natural fauna. Perhaps if we could identify some options then something could be done. It was a very simple idea.

I looked at sites from the north of England to the south, from the far east to the west. I spent many years with colleagues looking at sites. While some were genuinely non-starters, many others offered entirely suitable environments. When this result was reported back to their initially enthusiastic owners, it was generally then that the real problems began. Although some of the reasons given for inaction were credible, others were quite simply manufactured. As I listened time after time to excuses such as 'we have not achieved enough with our forest-planting policies to deserve them yet', or 'in one hundred years' time due to global warming this lagoon will be once again part of the English channel and therefore we can't reintroduce beavers here as we will have to remove them at some point in the future' being trotted out by the optimists, it became quite obvious that while big talk was easy, actual action was going to be hard to achieve.

Disappointingly, the nature conservation community, which is seldom a cohesive force, was not together at all on the beaver issue. Although some individuals in the statutory bodies were hugely supportive, others too ignorant, lazy or with contrary vested interests refused to help. They obstructed and undermined where they could. Beavers were a negotiating rag that could be thrown to opponents when they so wished to make limited gains on other issues. Other cheerleaders for rare aspen hoverflies, potamogetons and lichens feared that beaver felling or feeding would destroy the habitats of their cherished subjects. They refused to recognise that these species are adapted to what beavers do. Their blind bitterness caused confusion and delay.

In 2003, a conglomeration of organisations met at Newtown in Wales to discuss the prospect of reintroducing beavers. There was the normal grandstanding with windy speeches and carefree commitment. Reports. Papers. Conferences. Feasibility studies – seventeen years' worth of utterly wasted time, although a few ventures gathered pace and for a time offered tantalising hope.

———————

In 2008, South West Water proposed a project. Roadford Reservoir is a 295-hectare water body near the town of Launceston on the Cornwall/Devon border. It affords the strategic drinking water supply for tens of thousands of homes from the south Devon coast to the north. As a result of the rich cocktail of fertilisers and silts leaching from the farmed environment into the reservoir, toxic blue-green algae was becoming increasingly abundant. It's expensive to remove from the water supply at low levels, and if a bloom occurred in the warm summer months, the release of water into the River Tamar for the cleaning station at Gunnislake approximately twenty miles away would result in the death of pretty much all the river's life. An event of this sort incurs severe financial penalties for the water company and could potentially result in the company's chief executive going to jail. As this individual was at the time unsurprisingly unkeen on the last option, a solution to this issue had to be identified.

Martin Ross was the senior environmental engineer charged with identifying a solution. This small, calm, determined man had already pioneered the idea of upstream thinking whereby the company paid farmers to cover their manure stores, to impound upland ditch systems to re-established mires or to fence livestock out of river corridors to protect their banks in an effort to sustainably improve water quality. When he attended a conference at University of Exeter that explained how beaver-dam systems create natural filter systems, he became convinced that beavers could provide a solution to his problem.

He asked for my help and I surveyed the reservoir's entire upper catchment. There was plenty of perfect beaver habitat. Many kilometers of it. Around the reservoir's edge was much, much more. The initial consultations with the farmers who owned the land went well, as the prospect of being paid to hold water in unfarmable field corners appealed greatly to them. Field trips were arranged for groups of farmers to go to see beaver habitats in Bavaria and Brittany.

It was an education.

The farmers broke doors and drank bars dry. Pulled plug sockets out of the wall and pushed each other into the rivers. One late night in a Breton bar, they decided to play a drinking game called the 'Knights of the Flaming Arsehole', which operates on the basis of your pals choosing a drink for you – pints of beer were the most common option – and betting you could not drink it while lying face down on the floor with a rolled-up beer matt stuck up your bum, the opposite end of which had been primed with brandy. You were a winner if you drank it before the flames reached your flesh, and presumably a sore loser if not.

Thanking the Lord for traditional French values and a lack of any significant appetite for change, we used the soda siphons on the bar to douse the game players' flames before any 'rings of fire' became a reality.

A Disabled Walrus Might Be Possible

In Bavaria the farmers sympathised with the plight of the leader of the Bavarian Farmers Union, Thomas Obster, who turned out to meet them in a new Mercedes to explain how poor he was. He was the only person they actually listened to. Having visited Devon, Obster told them quite openly that 'if it were my land, I would release beavers onto it. They are fascinating creatures which do a lot of good. In our cornlands of Bavaria. In the low-lying valleys which depend on irrigation ditches, they are just the wrong species in the wrong place. Over there in your green-land, they will cause little issue for anyone. It would be a good place for them to be.'

Those attending came back having thoroughly enjoyed their free break with an entirely favourable view of beavers. When Gerhard came to visit a short time later, they extended every courtesy they could towards him. But opponents had rallied. A very few individuals sought to destroy its prospects for personal reasons. They held meetings in church halls, threatened milder neighbours and invented reams of rubbish so utterly ridiculous that they were hard to rationally counteract. One of the most cretinous was that as the Elizabethans had listed beavers as vermin, they knew something we did not. Although none of the cud chewers who listened spoke out, the Elizabethans let's not forget were the people who burned witches quite cheerfully and had internal church conversations about just how exactly missionaries were going to convert the tribes of dog-headed people they expected to encounter in South

America to Christianity when you said hi and they said woof in response.

They won when the water company's public relations team felt that the press coverage was becoming broadly unfavourable and drowning any other messages the company wished to advance.

———

In 2009, Sir Charlie Burrell called. This amazing man, who together with his wife Isabella Tree have rewilded with spectacular success their once intensively farmed family estate in Sussex to create a landscape that pulses with wildlife, had heard about my interest in beavers and was keen to help. He saw the arguments raised both for and against the return of the species, had attended a tour on Bavarian beaver management and while entirely supportive of their restoration from a personal perspective, felt that the polarisation of extreme viewpoints amongst interest groups was unhelpful. Charlie wondered if talking might help. He convened an informal workshop in his splendid home at Knepp Castle and invited representatives from pretty much every interest group there was to attend an initial meeting of what became the Beaver Advisory Committee for England (BACE). Most did.

Together with colleagues, during dinner I met representatives of the National Farmers' Union, the Country Land & Business Association Limited and various angling groups. In the past many of these organisations had been downright hostile or at least jolly unhelpful. As we sat long into the first night discussing the realities of beavers, it became quite clearly apparent that some of their concerns regarding, for example, the species becoming completely protected in law rendering their future management ponderous, were entirely valid. Equally, as we shared gin and jokes, they began I hope to realise that we were realists and, while utterly committed to restoring beavers, understood that a future ability to manage both them and their activities was quite critical.

Friendships formed that endure to this day. We still do not always see entirely eye to eye, but at least we are able to talk. It was a huge step forward.

In England, restrained by Secretaries of State who remained firmly focused on making farmers happy, English Nature had evinced little enthusiasm. There was evidence of some change in 2008 when they commissioned a feasibility study, which amongst other conclusions indicated that there was an abundance of suitable habitat and that beavers would do much good from an ecological and water-management perspective. This independent assessment undertaken by Dr John Gurnell recommended 'trial' reintroductions into exactly the kind of locations we had sought.[1]

In 2011, Mevagissey, a small village on the south Cornish coast, suffered a devastating flood. In the rumpus that followed, when sustainable solutions were required, the then director of The Lost Gardens of Heligan, Peter

Stafford, was sympathetic to the idea that the prospect of releasing a beaver population into the valley of the Meva-gissy stream was worthy of pursuit. A wealth of evidence from both Europe and North America indicated that the complex dam systems beavers create, which in turn form absorbent wetlands full of the felled timber they discard, slow the flow of floodwaters. Studies of the landscape suggested that beavers could be a natural solution. The local farming family who owned the upper valley was entirely supportive. The authorities were as enthusiastic as they ever got, and most external organisations agreed that it would be a very safe site for a trial. Enthusiasm built, and a licence application to release was prepared.

The prospects of this project indeed looked so bright that then Secretary of State for the Environment Richard Benyon felt the need to clarify his position. The Angling Trust, whose senior members were tediously convinced that beaver dams would obstruct the passage of migra-tory game fish, had lobbied him. Even though there were no game fish in the stream or indeed any other nearby river systems, any beavers anywhere were the thin end of the wedge for them. Despite a complete lack of evidence or sense, they had his ear. Benyon told both DEFRA and English Nature that whatever they decided he would have the final say. The implications of his statement were clear: that whatever we did, we would not get a licence.

In any case in the end it did not matter as Heligan's landlord, dismayed by the debate, decided that it would not be. He planned to restore formal gardens rather than absorbent wetlands in the lower valley. Beavers were no longer an option.

In 2019, the village suffered another entirely predictable flood event. It is certain that they will continue to occur and that those at the receiving end will continue to suffer. In Britain this perversity of land-use purpose is becoming all too increasingly obvious as the national policies for our moor-burnt, deforested, compacted, silt slipping, deep-drained island continue to deliver profits and pleasure for a subsidised few and total devastation for a much larger majority of others.

In Scotland a political change brought hope. In 2007, the election of the Scottish National Party to minority government brought fresh thinking. At a beaver-restoration conference held to relaunch the prospect in 2008 of another application to return the species to Scotland, a senior advisor to the new administration made it clear that they were sympathetic. A partnership formed principally by the Royal Zoological Society of Scotland and the Scottish Wildlife Trust reapplied for what was pretty much the same licence application with some wording rearranged to mask the utter idiocy of its first refusal in 2007. It was approved.

In April 2009, a small group of Norwegian beavers I had quarantined from Norway were released to the appreciation of a discreet crowd of well-wishers into a series of lochs in the Knapdale Forest. While one had a fit, 'turned turtle' and was found dead the next morning, the others moved out into their surrounding environments, established pairs and began to breed.

CHAPTER SEVEN

Trials and Tributaries

R ÓISÍN WAS CRYING. WE KNEW SHE HAD BEEN AT A stag party the night before and assumed it was just the alcohol at work. When she explained we weren't flying, we insisted that all was fine and that whatever the problem was we could sort it out. We were wrong.

I first met Róisín when she was employed as the field operation manager for the Knapdale project. Initially she was unimpressive. Angry and distant, a small Irish Goth girl with dark hair and panda eyes. In the years that followed as circumstance forced collaboration, I witnessed her volcanic tempers and became aware of her ice-thin sense of self-esteem.

She was a star.

Completely committed.

She loved her work and the creatures it involved. She did and continues to really care. In no time at all she had organised what was a difficult project into one that was so effectively structured and policed that it would have provoked the ecstasy of any third-world American dictator. She's a uniquely able individual and a beaver hero for sure.

Together with Alicia Leow-Dyke (the Welsh Beaver Project Officer with Wildlife Trusts Wales) and Rebecca Northey, my brilliantly well-organised project manager, we were travelling to a conference in Russia to see the oldest of beaver farms. What we had taken for visas were actually invitations. None of our smarter colleagues attending had bothered to inform us of this. We pleaded in vain with the customer-care staff at Heathrow who told us that if it were any other country in the world that yes, things would be sortable and that yes, it would be okay. As it was Russia, it was neither sortable nor okay. If they put us on a plane without a visa, we would each be subject to a £100,000 fine on arrival in Moscow, and it was very likely that the authorities would impound the plane and not give it back. Ever.

The fact that we were in exactly the same situation as around fifty attendees for a human rights conference in St Petersburg, who were unhappily gathered around their suitcases bickering like a clan of lost and dishevelled penguins on an ice flow heading towards the equator, was of little comfort. At moments like this in life, circumstance requires someone to show leadership.

Rebecca did just that. Pretty and elegant in a young Elizabeth Taylor mould, she showed the steel required to decide a course of action. We would stay and sort it out. On that Sunday morning, through a sponsor of ours, she contacted a New Zealander lass who worked in a visa-issuing unit of a travel consultancy for workers in the

Russian oil industry. Yes, she was prepared to come into work; if we obtained new photos and could present the documents, she would produce directly to the embassy on Monday morning, yes, the embassy would process. We discovered that Toby – who had also fallen into the same trap of misunderstanding – had genteel family living in London. As we raced around filling in the required forms, he parked us with them. They were kind. In their company we attended concerts and visited trendy wine bars and the Tower of London where the chap in charge of the Ravens told us unprompted how glad he was to learn from the press that a small group of beavers had escaped to live free on the River Otter in Devon.

Eventually we flew. From London to Moscow and from Moscow to Voroshnev where the beaver farm was. We had never been to Russia before, and as we landed on the former military airfield, the landscape stretched out vast, dun and solemn into the distance. Tall tree hedges separated great arable field systems as we drove towards the Biosphere Reserves centre. Through the gaps in their lines were glimpses of far tawny grasslands. Clusters of small straggling settlements lined the road. Yellow heating ducts made from tubular iron piping extended between dwellings on struts, around corners and up and over the roads at no great height. As I'd experienced the same sort of apparatus in Romania – where we had been told not to touch them as on the odd occasion when they were actually working they could be flesh-meltingly hot – these brought back old memories.

While some houses were of more modern construction, others had strange roofs made from entirely shiny, irregular metal with odd random, faded markings. It turned out these were wrought from the fuselages of Luftwaffe planes shot down in the tank battles, which had raged through the region during World War II. Others of much greater age were made from wooden uprights with herringboned infills of logs and clay. Where their front doors were open, you could see into backyards beyond where poultry, pigs and dirty small kids in patterned 1970s T-shirts intermingled in play. Fathers and grandfathers sat smoking or drinking outside. In their neat productive gardens full of fruit trees and vegetables, big-shouldered women with headscarves and short nylon dresses in vivid floral designs hoed weeds with long rakes. Beaten-up vehicles of all sorts were parked everywhere at random.

The beaver farm that once supplied breeding stock for the fur trade had been all but demolished. While odd random buildings still lingered with their unhinged doors rotting and glass windows shattered, they were being reclaimed by an undergrowth of mosses and ferns. Everything else was new. New conference centre. New approach road. New lighting. New bedrooms. Not enough to accommodate all those who were staying but new nevertheless.

We did not have time to linger and appreciate facades as the three days' delay incurred by our visa misunderstanding meant that our presentation times were imminent. Toby had five minutes to prepare and shuffled through his somehow. I had an hour to wait. As eastern European conferences can be tedious affairs with a distinct absence of jocularity, Gerhard and I had decided we were going to do a history of beaver–human interactions as explained in cartoons of which there are many. There were beavers attempting to fell wind turbines, beavers on the covers of fake men's magazines pretending to look moody while announcing to their intended readership how they got 'hard wood', beavers on an ark slowly sinking while Mrs Noah told Mr Noah that he 'never listened to a bloody word she said' and that she had told him 'it was a mistake to take them on board.' Beavers of all sorts to make you laugh for every social and cultural peccadillo. Unrestrained in our enthusiasm, we had decided to accept no limits to our social exploration of these associations. This proved to be our undoing.

As I sped through my slides with their candid social observations, an expert English-to-Russian translator in the back booth kept pace with ease. The western Europeans mostly laughed while the chaps from the east grinned hard when they felt it appropriate. All was going as well as could be expected until we got to an incredibly good cartoon from a 2005 edition of *The Sun*. The cartoon showed a man and a woman in a small car in a lay-by in the dark while two officials from DEFRA – labelled along the side of their van as the ministry of death – shone a torch at them and asked if they had seen any loose beavers. The woman was saying to her husband, boyfriend, partner or other sort of male acquaintance 'quick roll up the windows I have heard about men like that' and the point of the whole was that one could chart the presence of beaver humour even in the rising social trend for seedy people to go and have sex in the slime in a car park with complete strangers in the dark. It's called 'dogging'.

The problem was that even though it no doubt occurs in Russia, the context was so confusing that the translator had to stop and ask a rather unwilling Róisín sitting in the back row to explain. Róisín likes to wear cardigans and when she gets flustered or upset she pulls them round her ample front much like a broody chicken does with its feathers when you lift it off its cherished eggs to view them and it grumblingly repositions itself on top again once released. As the by now startled-silent translator listened with obvious incredulity to her explanation, she pulled her black hand knit forward and back so rapidly that the kinetic energy it was generating could easily have resulted in its becoming a fire hazard. When he finally understood her explanation, he refused to translate, which somewhat sank the continuity of my presentation as bewilderment spread through the substantially non-English speaking audience. Although most watchers were sympathetic, it ended with a whimper that was impossible to disguise. There is such a thing in life as a joke too far.

The following day we were shown round the beaver farm. It was small and indoors, with large swimming tanks and green-painted wrought iron bars to separate the contained beaver families from each other. There were both the black and normal nut-brown Eurasian colour forms in addition to a pair of burgundy Canadians with cinnamon faces and flank flashings. Although all the beavers looked well and there were babies, you could not help but feel that the facility had run out of purpose. Set up as it had been in the time of the early Soviet Government when beavers were very rare, at its peak Voroshnev had supplied thousands of individuals for reintroduction projects throughout the then Russian Empire.

Voroshnev had been hugely successful in assisting the reestablishment of the species. Its scientists had contributed greatly to the understanding of beaver biology and ecology and as an aside had fun breeding unusual colour forms for a laugh. As none were still living, the director showed us white beavers that were stuffed and photos of grey ones with white sides and white ones with grey sides. Quite what their function was God only knew, but it obviously made the old man happy and nostalgic.

The problem with projects of this sort is always: what do you use them for when their purpose is fulfilled and their course run? This obvious question had been solved, and to our amazement we were introduced by the Russians to a new and hitherto unplumbed use for beavers: they were being bred for a special circus.

Now, regardless of your views on wild animals in circuses, it must be quite obvious to most readers that beavers are not agile. They are not the kind of mammal that is going to jump through flaming hoops, leap elegantly onto pedestals, roar, ride a unicycle or perform a trapeze act in sequined spandex. From a biological perspective, all of this is quite impossible. But while beavers' talent for the showbiz world is limited, as we watched the near-incredible video they showed us it became apparent that they were broader than we thought. Beavers could, for example, slide down a small chute into a tub of water at the bottom (if you lifted them up to the top and then pushed). They could sit, or rather clutch, onto the back of a slow-moving Shetland pony without falling off (if you held onto them firmly). Finally, and most impressively, a beaver could play the cymbals wearing a red guardsman's tunic and a tiny black hat while appearing to march in an odd sort of way. When I asked Gerhard how on earth he thought it had been trained, he said, 'It's easy. You just sit them on a cooker and turn the back rings on alternately. They will get the idea in no time.' It was Russia. He might well have been right.

———

The colour of beaver protagonists is exquisite. Toby, who we met in the prologue, is a good example. Long a supporter of beavers and their reintroduction, he and Lavinia formed the Scottish Wild Beaver Group in 2011, initially to protest against the recapture of the beavers from the Tay. Now the organisation's lively website offers photo galleries, advice on how to survey for beavers, a wealth of information on the species' biology and a broad range of opportunities for seasonal events. They even organise frisky cultural events including plays where beavers are goodies and the baddies are. . . . Well, you know that answer by now.

The four children in the family are all accomplished thespians, journalists or political activists who variously provide their parents with moments of exquisite pride by getting arrested for their beliefs at various protest events.

Speaking of the theatre, Toby can become somewhat flustered during moments of high drama. One such event occurred in the summer of 2005 when my phone rang and on the other end was a breathless Toby.

'Derek, Derek', he panted. 'I am standing outside the village bakery and am so, so angry. I just thought I would call you to see what you thought I should do.'

Fearing the worst, I enquired as to the nature of his current calamity.

'Well it is really rather simple', he said, calming slightly. 'I had gone into the bakers to get some scones – the currant ones there are really rather good – when who should I see but my next-door neighbour Julian who told me rather curtly that he had a beaver building a lodge on his land and that he was going to shoot it! Well I was so angry I dropped my scones on the floor and, Derek, said the very first thing that came into my mind.'

With a sinking sort of feeling, I enquired as to what that had been.

'Well,' said Toby drawing breath, 'I called him a cunt. What do you think I should do now?'

Well I knew that Julian was a big landowner, and I knew that individuals of his sort commonly carry a lot of metaphorical as well as probable physical weight. It was likely therefore that he was well connected. As the beavers on the Tay at that time were below the radar of the authorities (who were focused on their official beaver trial in the west), it was all too predictable that if discovered they would throw every toy out of their pram at once.

'Apologise Toby, that's what I think you should do', I said. 'Go back in there right now, buy the man some scones and tell him how very, very sorry you are. Tell him perhaps that you are having a bad day. Tell him it's a toothache?'

'Well I really don't think, Derek, that I can do that', he said. 'No, I don't think that's possible; it's not really do-able at all. I am afraid I shall just have to go home now. Thank you. Goodbye.'

———

Chris Jones is another colourful example of a Beaver Believer, if a bit more 'happy clappy'. Imagine a short, stocky, bald man with a face that looks like a smiling emoji. Almost human. If the colour was right. Jones has led a life of great interest. As a policeman in the dying days of Ian Smith's UDI in Rhodesia, he has been on the run for years. If Interpol are still interested, I can provide all his contact details and put a sleeping draught in some of his award-winning hogs puddings. Any appropriate reward thereafter is mine.

Jones is also a visionary, an organic farmer who clearly can see the desperate need for a reintegration of agricultural best practices with nature's needs. He is passionate about soil restoration, agroforestry and drinking bottles of vodka through the spout of a teapot at the end of a night spent partying. I have the photographs to prove this in the event he denies.

Chris wanted beavers on his farm where he felt they would do a lot of environmental good and additionally slow the flow of his canalised Victorian watercourses, which were providing a more than sufficient supply of water to flood the Village of Ladock – which lies below his farm – on a regular basis in the winter months.

I organised them for him. At different times when others have a surplus, we hold in our farm buildings individuals singly as they await new mates. Chris built his enclosure with the assistance of Cornwall Wildlife Trust and, as we had a fair few spare females that I had already sexed, I arranged to pick up a young male on the way back from a work trip to northern England from a colony nearby. Chris had arranged a press call and a large party for the following day. Although the timing was tight, all I had to do was drive down past Hadrian's Wall in the far, far north back to sleepy rural Devon in the south, pick up a male in the late afternoon and then head down to his.

As a precaution before we introduce beavers to one another, we always check their sexes as a safeguard because mistakes on that front can be fatal, so I called Rebecca and told her just to pick a beaver in the shed, any beaver, check it was a female and that would suffice. She called back fifteen minutes later to say that as the first she had checked was a male, and as she knew a female was required, she had checked all the others and their sex was the same. Now as I had checked them myself, I knew I was right and that therefore by definition she was not. I told her to check again. Thirty-five minutes later she called back to say that she had and

143

that they were all males but as I knew I was right and was certain, I snarlingly told her to do so again. She did with the same result. As I sped farther south to do the sodding job myself and show them just how bloody wrong they were, a nagging spark of doubt crept into my thinking. Could I have sexed the whole damn lot and put down female instead of male? What I had moments earlier believed to be impossible was beginning to look rather likely.

What on earth was I going to do?

Cancelling was not an option. Chris had done his pre-TV interviews and all concerned knew that there were two beavers coming. We had no spare females. I could not think of any others anywhere. Then I remembered that, although hundreds of miles off my route, there was a young beaver in the stable of a pal of ours, surplus to his wants, in Wales. I phoned him. Did he know the sex? Hurrah, it was a female! Thank all that is holy I was coming now.

I altered my sat nav, added 400 miles to my journey and later that evening shook Keith's hand in the yard before we set off to get the beaver. It was in a stable with a deep straw bed. We cleared some of the straw and expected it to emerge. It did not. We cleared most of the straw and still no beaver. Looking at the last heap in the corner we figured it had to be there. We got our catching nets ready as we knew it was confined to a very small space and would come out fast, and gingerly started to remove the last of its bedding.

What we found was a very large hole. Gnawed right through the stable wall. Through this void we could see, glowing orange in the sun's last fading flare, the ancient glacial lake that the beaver had obviously headed straight for. We were stuffed.

I called Róisín and discussed putting two brothers long separated, out for the TV before catching them again rapidly in case they committed fratricide. No one could tell their sex, and providing each went to an end of Chris's very long lake for a short time, who would be any the wiser? It was a desperate solution but the stakes were high.

In the end Keith hit on the best possible solution. His beavers were tame.* They came into a feeding pen. He would sit through the night and close the gate when they came and bring us some 2-year olds in the morning. I drove home.

At 4.30 am he called to say he had three in their crates. When they arrived at my farm we sexed them and one was a female. No spectators for the happy beaver day that followed were ever any the wiser. Until now.

---

* I have never enjoyed the company of a tame beaver. Amongst the many we have moved, I can only ever remember one that would readily leave its lodge when I called. If I sat in his pen, he would bounce towards me with his tail wiggling, and when I leant towards him and slapped my hand on the floor, he would bound backwards gurgling happily then come right back for more. During the summer I kept him, I had little time to indulge his good humour and obvious willingness to play, although I can't remember whatever else was so important now. The other tasks that kept me so distracted. I regret now that I did not put them to one side to make pals with him. It would have been a better thing to do.

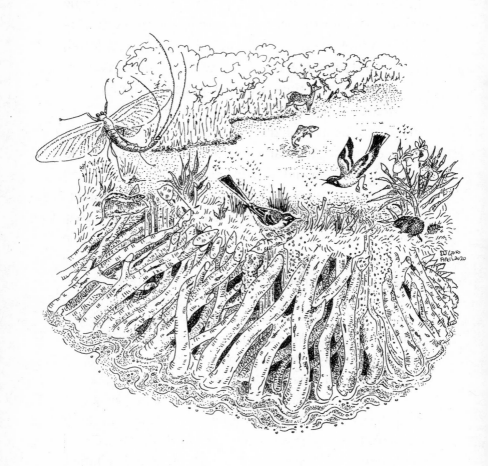

CHAPTER EIGHT

Beavers on the Otter

B EAVERS ARE BEING SLOWLY RETURNED TO BRITAIN.
But for all the jubilation generated in the media
and the press offices of organisations that are
keen to promote their own case, it has not been a professional undertaking.

Most workers in nature conservation will have heard,
for example, of an organisation called the International
Union for the Conservation of Nature (IUCN). Many
years ago, this worthy world body, stuffed full to the
gills of pretty sound experts, produced a recognised set
of guidelines to improve the way lost species are reintroduced back into habitats from which they had gone
– generally as a result of human misdemeanour. Those
of you just starting a career in nature conservation may
well have been taught by be-sandaled, beardy professors
with large cardigans to treat their strictures as command-
ments. But although a dwindling band of naive disciples
may still believe in the IUCN's worth, there are those
who employ their recommendations in a different way.

To use them as blocking mechanisms to achieve an end they value above all others: inertia.

It is unarguably true that without the actions of individuals who 'just did it' by accident or design, the return of the beaver in Britain would today be no more than a pipe dream. No nearer than a litany of tedious feasibility studies, action plans and witless, pointless computer models. Nothing tangible. Nothing real. Nothing really of any consequence at all.

If you think this absurd, then look at the recovery of wild boar – which are flourishing without any unnecessary paperwork or a permit to do so. They just escaped from poorly fenced farms, met others that had done the same and bred. If the restoration through official process of this species, which is one of the most significant agricultural crop pests in Europe, had been pursued in any other way, then it would quite simply never have been allowed to happen. The return of the hairy, tusky piggies would have remained a flight of fancy, which their computer modelers could have squadron led.

Between 2009 and 2014 in Scotland, the official beaver project resulted in the eventual release of a trial population of around sixteen individuals in the difficult landscape of Kintyre. Difficult because the choice of the site was designed to placate interest groups and opponents who were, are and always will be implacable. Difficult because it was bureaucratically over-burdened by hesitation and caution. Difficult not because there was no abundance of vegetation or good fresh water but because the landscape – with its lochans lying in the valley bottoms – was divided by high ridgelines that the beavers could not cross. Difficult for the young beavers born there to meet other beavers, establish new territories and expand their range.

Just difficult.

This site performed its function well. Its geography limited the extent and rate of the population's expansion. By the time it finished, pretty much the same number remained as had been released. It was not growing. Although other individuals from the River Tay and zoos are now being added, this site is in all probability not well enough connected to large landscapes to afford any long-term hope of viability. Politically acceptable, yes. Practically doomed, probably.

In contrast to Kintyre, escaped beavers from sources unknown expanded out from the River Tay to colonise every main stem of the longest river system in Scotland. It was easy for them to follow the river to find mates elsewhere, and the available habitat with its friable grassy banks, abundant wet-woodland and offline lochans was splendid. In 2009, kits were filmed playing outside a lodge on the water of Dean in Perthshire. While this may have been the first actual evidence of beavers breeding in the wild, it was quite obvious to those of us who were aware of their presence that they had been doing so for some time. In June 2010, an article in *The Scotsman* suggested that there was a population of between fifty and one hundred individuals living wild.

In direct response to this, SNH decided that these interlopers – which were so successfully unofficial – should be
culled. Their presence was embarrassing while Knapdale
stagnated. They were uncontrollable and could not be
contained. Against a mounting background of dissent,
SNH tried hard to justify their position by arguing that
the beavers might be diseased, Canadian or perhaps most
baffling of all that their welfare – by being allowed to
live and breed naturally in habitats they had chosen for
themselves that were near ideal – might be compromised.

When culling proved swiftly to be socially repulsive,
a strategy to capture them for internment in zoos was
proposed. It was quite obvious though that this was a
non-starter of a prospect. Beavers do not make good zoo
exhibitions and very few institutions wanted to keep them.
Those that did had captive beavers already. There was simply no space. Realistically it was a dim and desperate idea.

In the end the part-time trapper they employed with a
few traps, who had never before seen a beaver, caught a
sub-adult female in a fox trap in the snow. So inept was
his practice that one of her toes had to be amputated
in the days that followed her capture. As she had been
living on the River Ericht, she was nicknamed Erica and
was housed on her own in a concrete pen at Edinburgh
Zoo. As confusion reigned regarding her fate, some junior
individual in the zoo's marketing department accidentally
allowed her to be 'adopted' by the Alyth group of Beaver
Scouts. The zoo's conservation department, who were
unwillingly collaborating with SNH's policy of removal,

only learnt that their comrades had agreed to this when the *Daily Record* headlined the Scout's pleasure on the 10th of March, 2011. They were not pleased. When the kids enquired about her whereabouts and asked if they could come to see her, the zoo told them that 'she may be in Knapdale, the location of the official beaver trial re-introduction'. When it became apparent that she was not, they dismissed this statement as a 'clerical error'.

By the 7th of April, the *Daily Record* reported she was dead. Keepers at the zoo expressed their sadness 'to find the wild beaver captured by SNH dead in its enclosure.' SNH expressed its disappointment that the beaver had died at the zoo. It was not a honeymoon moment, and against a rising tide of disapproval, the trapping programme was quietly abandoned two days later.

After much grumbling, the authorities incorporated the Tay beavers with the small population in Knapdale under a single escutcheon: the Scottish Beaver Trial. Studies were undertaken of their distribution and numbers, and worthy committees pondered the beavers' fate. In the end, in May 2019, both groups of beavers were awarded the accolade of 'right to remain' by the Scottish Executive and thus became protected in law.

This was, however, protection with small-print caveats. There would be no more wild releases in Scotland and any other populations that appeared elsewhere would be removed.

A tiny population on the River Beauly in the north paid the price for a strategy approved by the Environment

Minister Roseanna Cunningham to assuage the ire of peeved potato producers. Although there were no root vegetables anywhere near the living space of the five that were captured, three died quickly in captivity later. Why?

Although populations now exist on other Scottish river systems such as the Forth, the current official policy is that only beavers that 'naturally disperse' will be allowed to form new populations. SNH, after a sound metaphorical 'tasering' from agricultural ministers, has agreed that this is reasonable and that as beavers now exist in Scotland in a population that demonstrates 'good conservation status' (at a current estimate of approximately 500) that this species, although it is protected throughout the European Union, should be shot when conflicts arise with landowners who dislike them and resent their return, rather than be moved to areas of suitable habitat elsewhere.

As totems of change.

Of a challenge to their hegemony.

As an expression. Living and breathing of the point of view of others.

Although SNH points out that the lethal control of beavers is a legitimate population-management tool in other states such as Bavaria – which is approximately the same size as Scotland – Bavaria accommodates a population of 22,000 beavers in a farmed landscape that is much more intensively used. In Bavaria, beavers live through every major watercourse. They are tolerated in farmlands where possible and, where they cannot be endured, managed pragmatically through a system organised in part by the state government. They have become the new 'normal' accepted by many people in villages, towns and cities. Where there is no option for surplus, which for a decade beaver managers avoided by shipping thousands out free of charge to vacant habitats in countries such as Croatia, Romania, Hungary and Belgium, they will kill. Reluctantly, without rancour or hate, and under licence by approved beaver managers.

The situation is not the same as Scotland, where now after a single season of official licensed culling in the year 2019/20 it is believed that slightly under one hundred individuals may have been killed. If this figure alone is correct – regardless of any other clandestine killing – then Scotland could be heading for a *Guinness Book* award as the first nation in 'civilised' history to both reintroduce and then extirpate a beaver population in the shortest possible period of time.

Society does not want this. People are disturbed. Children have written illustrated reams of poetry to the country's first minister to beg her to stop the killers of eagles, harriers, badgers and beavers in their tracks. Still it continues.

While insiders say that SNH was unwilling to follow this current strategy, it has been either persuaded or told to do so by politicians. When it comes to most issues of contention involving nature conservation in our modern world, the views of the other, louder land-use organisations commonly reign supreme, however illogical.

Brash, bearded, big men stand on stage and tell in tomes of thunder how their members (who are of course all conservationists) will not tolerate the loss of one salmon, one lamb, one pheasant, one grouse or one carrot. Their point of view is, they believe within their closed ranks and minds, entirely reasonable.

They are a settler culture. If this seems absurd, any student of history will tell you that it was always so. Strident voices on remote frontiers demanding of timid mandarins and weak politicians extreme actions against both humans and non-humans to support their own selfish ends. The First Peoples of North America were decimated in this way. The Congolese the same. The thylacines the same. The Pieman River people, California grizzly, grey wolf, Zulu and bushmen. All the same. Beavers are only the most modern of victims standing in line on the temple steps of ignorance to be sacrificed by the primitive priests of progress to their great god of greed.

There is no conceivable reason why this lobby should be placed in a position to advise, dictate or influence an environmental policy to which they are uniquely opposed. The death penalties now being delivered for simply being a beaver have been enacted with no opportunity for discussion in wider society.

None at all.

No consultation with children whose world they would positively impact.

On the River Tay it is likely that many more beavers are in any case being illegally killed, in addition to those that have been licensed to die. There is plenty of evidence in the form of burnt-out lodges and bloated, bullet-ridden cadavers. No culprits have ever been found.

————

In February 2014, an amateur wildlife cameraman named Tom Buckley filmed two beavers playing in the River Otter in Devon while a third fed on a tree behind them. Although there had been sightings of a single beaver on the river for years – a supposed escapee from a local wildlife centre – no one believed there to be any more, but on his infrared trail cam was a family living wild in England. The press interest was significant, instant and completely positive.

The farming family whose land they were living on was wonderful. Pleasant, intelligent and kind. David and Vicki Lawrence were young. A new farming generation with a different view of how the landscape might be. Vicki had established a glamping safari tent business on their modern 250-acre dairy unit. They were delighted to provide the beavers with a home.

Those of us wishing to restore them to England had made no further progress. But this story was amazing. Breathtakingly hopeful. Positive and pure. I went to see Tom, who by that time had assembled more footage. It was obvious even then that the three were not alone; field signs, gnawed sticks and felled trees were also present elsewhere. Great news. Tom, a calm, retired scientist from the Environment Agency was concerned about what was going to happen next and rightly so. It was obvious that either the government would have to say yes to their staying on the land. Or no.

Owen Paterson was the Secretary of State for the Environment at that time. He was no friend of nature. As a firm supporter of farming and shooting interests with little regard for conservation at all, the receipt of a single letter from the Angling Trust whose knowledge of beavers remained alphabetic decided his mind. They would be destroyed. As he was the politician who famously blamed the badgers for 'moving the goal posts' when they simply refused to die in the numbers required for a dubious bovine TB elimination cull, the response to his statement was predictable. Near complete dissent. Within a week his position, buffeted by an entirely negative media and social response, changed to 'no, they cannot stay, but must be removed to zoos'.

Although this script was near identical to what happened with the Tay, the position of other organisations that might have been inclined initially to support his thinking had begun to shift. As media and public support for the beaver's retention rose, a whole host of allies began to emerge. The Devon Wildlife Trust became the public face of a coalition of the willing. Of advocates for change.

Together with pals I attended initial meetings with senior civil servants from DEFRA. They insisted that no third way was possible before asking with an airy arrogance if we would like to cooperate with them on the capture-and-zoo plan by providing traps and expertise. We said no. I spoke to some of the zoo directors who had been asked to provide future homes. I explained the issue at stake: that no wild beavers on the River Otter meant no wild beavers perhaps ever in England. They all said no to DEFRA as well.

One of the most surprising moments for me personally was when Pete Burgess from Devon Wildlife Trust and I attended a meeting with two representatives from the National Farmers' Union. It was refreshing. One was a friend, a clever, sympathetic man whom I had met through Charlie Burrell's Beaver Advisory Committee for England meeting and had come to admire very much. The other who opened the meeting I did not know. It was this individual who provided food for thought. He made it clear that while they were there to present the official party line, which was of course 'no, the beavers cannot stay', that this was not their view as individuals, that they considered the location to be as good as any could be for an English beaver trial and that furthermore, as the political landscape of land-use was obviously shifting even then, that they valued the working relationship they had established with us and had no wish to see it jeopardised. We told them we felt the same. They suggested therefore that while we were on both sides going to have to toe a party line, would it be possible to collaborate where we could and to ensure that when achievable we sprang no surprises on each other by discussing issues in advance? That way we could soften any impacts or hiccups to avoid needless acrimony and to ensure that any future debate was as rational as it could be.

During the months to come they attended incognito some of the public meetings and saw the many people from different backgrounds who spoke in support of the beavers. They listened and helped when they could and did not support the government's failing public line. We were terribly grateful.

Chris Price, the then head of environmental policy for the Country Land & Business Association Limited, a body that had been very difficult in the past, expressed his employer's current position. Beavers were cool. Most welcome. Absolutely spot on. Although it was unlikely that all of his more traditional colleagues quite shared his enthusiasm – or even understood the 'cool' thing – Chris, who is a look-alike candidate for the position of Brian Blessed's bucolic, stouter but much, much younger brother, carried the day. It was fantastic news.

All DEFRA had left were the Angling Trust, who were worse than useless. They asked David Lawrence if they could film near his beavers for a TV piece on how beaver dams would impede the migration of game fish – a nonsense supported by as much evidence as there is for pink unicorns with horns made from honey – and he said no, too. Other more cerebral fisheries organisations considered a beaver trial worthy, or at least did not want to look dim by opposing something that so clearly had science on its side.

Together with colleagues from the Devon Wildlife Trust, I was offered the opportunity to meet junior environment minister Lord de Mauley to discuss the issue further.

On a warm summer day, we travelled to see him in his dark panelled office in Westminster. An ex-military sort, he was charming and kind, and although he told some rather odd jokes about bearskin hats, we were offered some nice tea and shortbread. He asked how he could help as he really knew nothing of beavers and would like to resolve any impasse amicably. We agreed, as he was of course a busy man who could not be expected to understand every scintilla of his wide portfolio, that his lack of knowledge was quite reasonable. But when we put our case that beavers should be allowed to stay, his civil servants on the opposite side of the table insisted that they would not. Lord de Mauley suddenly remembered a host of deep-seated but rather vague concerns, which he was unable to outline in detail. Despite assuring us he had no knowledge when we entered, he told us just how concerned he was when we left. We made no progress.

Approximately 70,000 people signed a petition to demand that the beavers remain. Hundreds of letters were sent to the Ministry. The Friends of the Earth served a freedom of information request that illustrated that, as DEFRA had no idea of what they were talking about and no one to advise them, every individual query was stamped by DEFRA's grey little clerks with the same 'standard response'. The beavers were illegal immigrants that might be diseased and although they had bred and done no perceptible harm, the habitat they were living in might be unsuitable. All the normal guff. For ordinary people whatever their question. The single exception was the wealthy philanthropist and landowner Lisbet Rausing, who got exactly the same response but in handwritten, fawning form from the minister himself.

The Animal and Plant Health Agency (APHA), DEFRA's 'Milice', were told to catch and detain the beavers. More public meetings were organised in Ottery St Mary and Honiton. The news media and hundreds of people attended them. One teenage girl gave the most moving presentation about nature conservation I have ever heard in my life, about her wish for a better future of recovering nature inspired by the beavers that were living at the bottom of her grandmother's garden. She, more than any other, epitomised the reality that their skills were vital to cauterise the hemorrhage of life from our land. To heal and repair. An 11th-hour solution at the point when saviours are sought. When she finished, people clapped and cheered.

Parish councillors, car salesmen and shopkeepers agreed. Farmers spoke in support of beavers' retention. Voices never heard, never listened to in the normal secrecy of land-use debate sounded strident and clear. The BBC news told us that APHA were that day going to Scotland to uplift the beaver traps offered by SNH.

No one was pleased.

Elegant politicians leant their effusive support when young and old, poor and rich, landed and landless combined to say no. Then to jubilation when Owen Paterson was sacked and a new environment minister appointed, Devon Wildlife Trust was awarded a licence for a five-year study trial to begin. The beavers could stay.

It was over.

───────

Five years later, the Wildlife Trust's study, too, is now nearly done and it demonstrates quite graphically that beavers afford all they promise on the box. That they increase desperately depleted nature in every way possible in great wealth and abundance. That they naturally in their created landscapes slow flows, and retain and clean water. That they are a critical missing component of our nature and that we must have them back once again everywhere they can be.

───────

Soon a new minister will make a decision regarding the national future of beavers in England. Will it be restoration fast on a national scale or more witless prevarication and delay? Will big, brash, bounding Boris Johnson leap to deliver beavers with his blond mane bobbing in the same wham-bam style as he accomplished with Brexit? Or will the so much tireder, more contemplative figure recovering from his near death experience with coronavirus take time to think again about reforming a much less conventional, regenerative new world when this human plague comes to its end?

Will he care about where we are now?

About a future world for our children?

It does not have to be the way it was.

We can change, it is possible and would be better if we did.

Right now the early indications are that this may be so. Although we cannot meet, the powers that be are giving every indication that it's possible. That politicians want to collaborate, that they see the merit now and are moving to a position of support for the return of the beaver to Britain.

As I finish writing this book on a sun-strobed porch overlooking a green, rolling agricultural landscape of pastures, hedges and scattered oak woodlands, I know that down deep in the valley beneath me there are beavers. On a gravel pit to the east is a lodge and on its exit stream many dams. When I drive past on the road above, I can see newly felled trees lying peeled white in its deep, clear water.

Its owners don't mind.

In the fading light of this spring day when the sun's warmth outside is drawing life up from the soil, I know that, soon as the evenings draw long, there are pools where I can go to sit still. To watch while the fish jump and the birds bathe in the shallows, the beaver babies as they wrestle and play. I can hear them clearly as they murmur and squeal with delight.

Sometimes a fox stops to watch as well in contented company.

In England, Wales and Scotland, beavers are returning. Slowly. To the joy of many and the irrational ire of a remaining few. The beavers have many more supporters now than enemies. In all classes and walks of life people, children, politicians, landowners, farmers, fishermen, friends. Without government action to exterminate them on a landscape scale, their populations will expand untrammelled for sure. They will endure, and failure is done. The restoration of beavers to a river system near you – wherever you may be – is therefore no longer a case of uncertain *if* but now quite predictably *when*.

Welcome them when they come home. They are old friends.

Epilogue

IT'S EARLY SPRING IN A DEVON WOODLAND. YOU CAN smell the warmth – a pleasant, nut-chocolate – rising from the damp soil as its waking life pulls greedily down on the last of the dry autumn leaves. Paper dry or skeletal with their ribs alone intact, they will merge softly into the earth to become one.

The verdant green of the newly hatched buds is so brightly luminescent it hurts to focus on their elfin splendour. Bluebells flower in cobalt ranks. The soft butter-yellow of the primroses supersedes the stronger sundrops of the fading celandines. Where the wet advances, great globes of marsh marigold prepare to erupt in full bloom.

Fleshy orchids thrust forth dappled leaves. Their pastel pink wax flowers will soon burst forth. Wood anemones litter the floor like bright fallen stars. A bumblebee burrs past.

Birdsong is everywhere. Early nests of blue tit babies demand their gaping beaks be stuffed full to the brim. Hard-harassed, their painted parents bustle busily to achieve their ambition.

In the far distance, a grasshopper warbler high in a willow clearly and repetitively chirrs the call of its summer namesake. It will descend to the ground soon and scutter to its nest hidden dense, interwoven in the rank grass thatch.

Although a plantation of only twenty years old, in its hedges, banks and boundaries a few old oaks remain. As the rhythms of the new season burst forth from their seasonal slumber, not even the big trees are old enough to recognise the new beat of life.

In the dumb of our silence, we cannot hear.

Its throb is ancient and strong.

The land knows its lilt.

In a growing cacophony of gnawing, crashing, splashing and mewing, beavers are back.

Where they return, gnawed sticks with flute-shaped ends lie white, stripped of their bark, in tumbled clusters. Felled trees with pencil-sharpened stumps are silent in nascent clearings. Embankments raised by Victorian drainers become islands. Streams run back into courses they choose for themselves. Gullies worn down to bedrock become pools. Pools become ponds. Ponds become lakes and lagoons.

A multitude of them.

As the flow slows and spreads.

On land, toppled trees afford space for small mammals whose hidden seed stashes burst forth in new life. If they crash in the water, fish fly through their canopies, hiding hard from green-gobbling cormorants and sleek, swimming otters.

Where felled branches stand up, a flash of aquamarine and orange announces the piping presence of the fish feeder. Perched high, looking down, the kingfisher can see the minnows turn clear. Multitudes of them hiding in the fallen trees' tangle. Hiding from him. Shardlike he dives down and then up to bash dead his victim before searing off through a void in the hedge bank and into the dark.

Muddy paths ground by gigantic, webbed feet lap canals that twist through the wetting and into the fields beyond.

Where it suits them, when they bother, the beavers work hard to extend their detritus-based dams. Mud and roots. Stones and sticks. On occasion for fun, farm waste to amuse. A traffic cone. A tyre. Chemical sacks from the 1970s with ICI logos.

Water. Water in pools. Deep and dark when they have been for a time. Eerily translucent where the grasses and field flowers wave slowly in their last drowned dance of death in the bed of new impoundments. On the surface whirligig beetles like drops of spilt solder race and revolve. Their great diving cousins oar themselves down, nut-brown and burnished, to feed on dead worms.

Water flowing in broad blankets. Water slowly squelching. Raising the land to form smelly sphagnum sponge.

Sometimes they don't care and a wary youngster at dusk will stop munching as you pass close. It will follow you with its steady hazel eye 'til the white shows and you have gone. Then it will start to feed again, gnawing loud and rhythmic. Normally you will be fortunate to see them as their gigantically, furred forms slip languid and silent through the engineered safety of their deep, drunken kingdom. With their hippo heads flat on the surface they are ever alert for danger. As they circle 'round, they flick their heads back, drawing draughts of evening air through their large, fleshy noses.

They do not know that the bears, lynx and wolves are now all long gone and that the Mesolithic hunters who left their flint-knapped discards in the pastures above have also returned to the soil.

In a slimy moving mass, croaking common frogs spawn where they have not for untold generations. Clasping toads congregate to provide a seasonal bounty for predators. Tangled trails of black-dotted egg spawn weave past the spotted male newts, enticing their drab, unblinking partners with their dragon crests and violet tail bands.

Dragonflies whir through the tall herbs as their buds ready to flower. Delicate blue damsels and needlelike reds tremble on reed stems in the gentlest of gusts. An emperor with its green-enamelled head strikes a day flying moth on the wing. Its scales puff then slowly snow downwards.

Agrions, metallic in dark green or deep blue, shimmer and shine over water trickling clear in the full sun's light. Their dark banded wings distract your eyes' view.

Wary water frogs watch the golden-eyed grey herons stalking slowly in the shallows. In the weed-sheltered centre of the mere, they bubble-gum puff their throat sacs as a great, green grass snake undulates out towards them.

Sometimes a giant walks out of the woodland and pauses with his front leg held high to sniff the coming night's air. Burnished orange-ochre with high crowned head. Smooth, sleek and shining. A red stag, the last of the great beasts. He may stop to sniff the surface of the water, to watch the ripples of his breath force flow forward from his face. Dark-eyed he looks down at his own ear-flicking reflection as the horseflies attempt to draw hard on his blood.

Reintroduced water voles, replete and round, ignore their jumping offspring. Their ripe, rich piles of territorial droppings tell neighbours to beware of their pugnacious mothers, hunched inside the burrow entrances. They are always ready for a fight.

Moorhens and mallards, grey geese and woodcocks live their own lives. Roedeer return to feed on the tall grass and bramble that was once stripped bare by sheep.

Red foxes snooze, scratching in the sun.

Insects in their multitudes whir, chirr and thrum in abandon. When you close your eyes and relax, you can still hear the buzzing.

For trees the relationship is special. Once beaver-felled they coppice with ease, sending their smooth-skinned shoots skywards. In early, lithe life, they generate toxins to deter the deer browsing. Poplars take from cuttings where lodged in rich silt. Willows grow a meter in a single season when their seeds set in damp. Aspen sucker out from their felled mothers to form fragrant forests of trembling children. Some, ghoul like, re-root from dislocated limbs when their hoary rind splits to revive in unstoppable life.

Although many do drown, their cadavers can remain standing for years. Slim limbs rotting to powder and falling. Hard heartwood standing tall. Cosy bat roosts where old bark flakes up. In soft-wooded sockets, willow tits excavate while woodpeckers' boring and pneumatically pocking provide dwellings for squirrels and nook-nesting owls.

As they labour and burrow; create woody dams; scent mark their leaf piles with their camphor-rich juice; build, browse and bicker; and fell trees with ease, the beavers wrapped up in their works for a day, don't know, care or wonder that all that surrounds them is made by their actions.

That it needs them and without them would fade swiftly away.

— ACKNOWLEDGMENTS —

I WOULD LIKE TO THANK VARIOUSLY FOR THEIR CONsiderable encouragement, support and assistance during the very many years it has taken to restore, at least in small part, beavers to Britain: Gerhard Schwab, Martin Noble, Roy Dennis, Sarah Bridger, the late Kenneth West, Dr Bryony Coles, the late Dr Russell Coope, Sir John Lister-Kaye, Prof Alastair Driver, Paul and Louise Ramsay, Isabella Tree, Nigel Carter, Sir Charlie Burrell, John Smellie, Dr Róisín Campbell-Palmer, Dr Pat Morris, Ted Green, Dr Brian Thomson, David Anderson, Lord Goldsmith (of Richmond Park), Ben Goldsmith, Dr Grahame Bathe, John McAllister, Jonathan Spencer, Martin Ross.

— NOTES —

Chapter 1: The Salvation of St Felix

1. David Keys, 'Revealed: How Prehistoric "Des Res" Gave Stone Age Brits a Perfect Diet', *The Independent*, December 10, 2013, https://www.independent.co.uk/news/science /archaeology/revealed-how-prehistoric-des-res-gave -stone-age-brits-a-perfect-diet-8995918.html.
2. Bryony Coles, *Beavers in Britain's Past* (Oxford, UK: Oxbow Books, 2006), 55.
3. Rachel Poliquin, *Beaver* (London: Reaktion Books, 2015), 11.
4. Róisín Campbell-Palmer, et al., *The Eurasian Beaver* (Exeter: Pelagic Publishing, 2015), 14.
5. Volker Zahner, et al., *Des* Biber *Baumeistor mit Biss* (Regenstauf: SüdOst, 2020).
6. Jacques Boudet, *Man and Beast: A Visual History* (London: The Bodley Head, 1964), 176.
7. Giraldus Cambrensis, *The Historical Works of Giraldus Cambrensis,* ed. Thomas Wright (London: H.G. Bohn, 1863), 431.
8. Hūšang A'lam, 'Beaver', Encyclopædia Iranica, 15 December, 1989, http://www.iranicaonline.org/articles/beaver -castor-fiber-1.
9. Efraim Lev and Zohar Amar, *Practical Materia Medica of the Medieval Eastern Mediterranean According to the Cairo Genizah* (Leiden: Brill, 2008), 354–55.

10. Muhammed Ibin-Mūsā Ad-Damīrī, *Ad-Damīrī's Hayat al-Hayawan, Volume I*, trans., A.S.G. Jayakar (London: Luzac & Co., 1908), 481.
11. Chrysanthe Pantages, 'The Fur-Cloaked Prophet: Elijah and the "City of Beavers"', *The Iris*, 2 June, 2014, https://blogs.getty.edu/iris/the-fur-cloaked-prophet-elijah-and-the-city-of-beavers.
12. William Anthony Holmes-Walker, *Sixes & Sevens: A Short History of The Skinners' Company* (London: Skinners' Company, 2005), https://www.skinners.org.uk/historical-timeline.
13. Holmes-Walker, *Sixes & Sevens*.
14. Earl L. Hilfiker, *Beavers: Water, Wildlife and History* (Interlaken, New York: Windswept Press, 1991), 60.
15. Adrian Tanner, *Bringing Home Animals: Religious Ideology and Mode of Production of the Mistassini Cree Hunters* (New York: St Martin's Press, 1979).
16. William L. Sachse, 'England's "Black Tribunal": An Analysis of the Regicide Court', *The Journal of British Studies* 12, no. 2 (May 1973): 69–85, https://doi.org/10.1086/385642.
17. Ben Goldfarb, *Eager: The Surprising, Secret Life of Beavers and Why They Matter* (White River Junction, Vermont: Chelsea Green Publishing, 2018), 43.
18. Cambrensis, *The Historical Works*, 431.
19. Edward Topsell, *The History of Four-Footed Beasts and Serpents* (London: E. Cotes for G. Sawbridge, 1658), 37.
20. Jason G. Goldman, 'Once Upon a Time, The Catholic Church Decided that Beavers Were Fish', *Scientific American*, 23 May, 2013, https://blogs.scientificamerican.com/thoughtful-animal/once-upon-a-time-the-catholic-church-decided-that-beavers-were-fish.
21. Topsell, *Four-Footed Beasts*, 38.
22. Coles, *Beavers in Britain's Past*, 133.

Notes

23. Coles, *Beavers in Britain's Past*, 182.

24. *Ordnance Survey Name Books, Ross and Cromarty, 1848–1852*, vol. 47, 22, https://scotlandsplaces.gov.uk /digital-volumes/ordnance-survey-name-books/ross-and -cromarty-os-name-books-1848-1852/ross-and-cromarty -mainland-volume-47/22.

25. *Ordnance Survey Name Books*.

26. Coles, *Beavers in Britain's Past*, 191.

27. Coles, *Beavers in Britain's Past*, 182.

28. Cambrensis, *The Historical Works*, 429.

29. Cambrensis, *The Historical Works*, 430.

30. Gover et al. 'The Place-Names of Nottinghamshire', *EPNS* 17.

31. Leslie Peter Wenham, *Watermills* (London: Robert Hale Ltd, 1989), 150.

32. John D.A. Widdowson, 'Some West Country Lexical Elements in Newfoundland and Labrador English', *Transactions 149* (2017): 213–32; Coles, *Beavers in Britain's Past*, 222.

33. Eilert Ekwall, *The Concise Oxford Dictionary of English Place-Names*, 4th ed. (Oxford: Clarendon Press, 1959), 40.

34. Derek Gow and Coral Edgcumbe, 'A History of the White Stork in Britain', *British Wildlife* 27, no. 4 (2016): 230–38.

35. Coles, *Beavers in Britain's Past*, 180.

36. Coles, *Beavers in Britain's Past*, 187.

37. Coles, *Beavers in Britain's Past*, 147.

38. Edmund Bogg, *Higher Wharfeland: The Dale of Romance, From Ormscliffe to Cam Fell* (The Old Hall Press, 1989), 26.

39. John Clare, 'Remembrances' (1832), 78.

Chapter 2: Spurting Streams of Grease

1. Tansley, personal communication, 2019.

2. Peter D. Moore, 'Sprucing Up Beaver Meadows', *Nature* 400 (August 1999): 622–23.

Chapter 3: Popielno

1. Bryony Coles, *Beavers in Britain's Past* (Oxford, UK: Oxbow Books, 2006); Barry Driscoll, 'Bring Back the Beaver', *British Wildlife* 19, no. 11 (1977): 493–7.
2. R.S.R. Fitter, *The Ark in Our Midst* (London: Collins & Harvill, 1959), 105.
3. R. Loder, personal communication.
4. William Onslow, 'Why not a National Park in the Highlands?', *The Countryman* (January 1939): 496–507.
5. G.G. Stewart, personal communication, 2004.
6. Martin Noble, 2019, personal communication.
7. Driscoll, 'Bring Back the Beaver', 497.
8. Sir Christopher Lever, 2018, personal communication.
9. Horace T. Martin, *Castorologia. The History and Traditions of the Canadian Beaver* (London: Edward Stanford, 1892), 102.

Chapter 5: If You Look Now at Her Face

1. Duncan J. Halley and Frank Rosell, 'The Beaver's Reconquest of Eurasia: Status, Population Development and Management of a Conservation Success', *Mammal Review* 32, no. 3 (September 2002): 153–78, https://doi.org/10.1046/j.1365 -2907.2002.00106.x.
2. 'Beaver', Zoroastrian Kids Korner, last modified February 15, 2020, https://www.zoroastriankids.com/beaver.html
3. Edward Topsell, *The History of Four-Footed Beasts and Serpents* (London: E. Cotes for G. Sawbridge, 1658), 36.

Chapter 6: A Disabled Walrus Might Be Possible

1. John Gurnell, et al., *The Feasibility and Acceptability of Reintroducing the European Beaver to England*, Report Prepared for Natural England and The People's Trust For Endangered Species (March 2009): 75.

— INDEX —

Index

Index

Index

Index

Index

Index

— ABOUT THE AUTHOR —

D EREK GOW IS A FARMER AND NATURE CON-
servationist. Born in Dundee in 1965, he
left school when he was seventeen and
worked in agriculture for five years. Inspired by
the writing of Gerald Durrell, all of whose books he
has read – thoroughly – he jumped at the chance to
manage a European wildlife park in Central Scotland in the late 1990s before moving on to develop
two nature centres in England. He now lives with
his children, Maysie and Kyle, on a 300-acre farm
on the Devon/Cornwall border, which he is in the
process of rewilding. Derek has played a significant
role in the reintroduction of the Eurasian beaver,
the water vole and the white stork in England. He
is currently working on a reintroduction project for
the wildcat.